Postharvest Management

Postharvest Management

Editor: Elijah Baxter

R CALLISTO REFERENCE

www.callistoreference.com

Callisto Reference,
118-35 Queens Blvd., Suite 400,
Forest Hills, NY 11375, USA

Visit us on the World Wide Web at:
www.callistoreference.com

ISBN: 978-1-64116-189-3 (Hardback)

Cataloging-in-Publication Data

Postharvest management / edited by Elijah Baxter.
   p. cm.
Includes bibliographical references and index.
ISBN 978-1-64116-189-3
1. Crops--Postharvest technology. 2. Agricultural processing.
3. Agriculture--Management. I. Baxter, Elijah.
SB129 .P67 2019
631.582--dc23

Table of Contents

Preface

Postharvest handling is the process of cooling, sorting, packing and cleaning the crop after harvest. It is a stage of crop production in which the crop is removed from the parent plant or is removed from the ground. At this stage, the crop is vulnerable to quality deterioration. Postharvest treatment determines whether a crop is suitable for food processing or for fresh consumption. The targets of postharvest management are to keep the product cool and its moisture content intact. It also strives to prevent physical damage to delay spoilage. Sanitation is also an important aspect of postharvest management. High-speed cooling, refrigerated and atmosphere-controlled environments are used in large operations to prolong freshness. After harvesting, shelf life is dependent on the overall flavor, taste, texture and appearance. Postharvest technology is crucial for stimulating agricultural production, preventing postharvest losses and improving nutrition. This book provides comprehensive insights in the field of postharvest management. It strives to provide a fair idea about this discipline to help develop a better understanding of the latest advances within this field. Students, researchers, experts and all associated with postharvest management will benefit alike from this book.

The information shared in this book is based on empirical researches made by veterans in this field of study. The elaborative information provided in this book will help the readers further their scope of knowledge leading to advancements in this field.

Finally, I would like to thank my fellow researchers who gave constructive feedback and my family members who supported me at every step of my research.

Editor

Cauterizer Technology Increases Cactus Pear Shelf Life

Tebien Federico Hahn

Additional information is available at the end of the chapter

Abstract

Cactus pear is a food of nutraceutical and functional importance worldwide. Improvements in cactus pear shelf life will allow international 1-month refrigerated shipments for supplying good-quality fruit to the European markets. The perishable fruit is chilling sensitive and hot water treatments by immersion, increased shelf life to 1 month. Harvest tools are important to avoid fruit damage to the stem end, which leads to pathogens attack and fruit decay. A light wire-cutting tool detached 300 fruits h^{-1} without leaving the worker with pain. A hot cauterizer (HC) performed better than a cold cauterizer (CC) extending fruit shelf life to 2 months; 85% of the HC-processed fruits were marketable after 2-month storage at ambient temperature. After removal of the prickle pear peduncle, the pulp contacts a hot flat metal surface at 200°C. A pressure of 100 kPa during 30 s assures proper heat transfer and surface healing. An automated HC process-line cauterizes 250 kg h^{-1}.

Keywords: hot cauterizer, cactus pear shelf line, hot cauterizer line, automation

1. Introduction

Opuntia ficus-indica domesticated probably 9000 years ago [1] in central Mexico is consumed as fresh fruit since ancient times. Cactus pears of the Cactaceous family are endemic from America [2, 3], growing more than 3,000,000 ha of *Opuntias* in their native natural habitat [4]. Once introduced into Spain from Mexico, the species spread throughout the Mediterranean [5], and today, cacti are cultivated in southern Spain, France, Greece, Israel, Italy, and Turkey. The Arabs took them from southern Spain to the north of Africa being now cultivated in Algeria, Egypt, Morocco, and Tunisia [6]; Tunisia cultivates 400,000 ha [7].

More than 70% of all the *Opuntia* species grows in the arid and semiarid regions of Mexico, Argentina, Peru, and Chile [8]. The largest area for fruit production is found in Mexico, with an estimated cultivation area of 72,000 ha for prickle pear fruit growth and 10,500 ha for cladodes [9]. Peru has approximately 35,000 ha of wild plants used for cochineal breeding [5], while Brazil has 40,000 ha for forage. People living in Angola, Australia, India, and South Africa [10, 11] are looking for plant species that can adapt to arid and semiarid zones that can provide food and materials [6]. Italy cultivates 2500 ha for cactus pear production and Chile around 1100 ha.

1.1. Importance of prickle pear

The cultivation of prickly pear started as a resource for erosion control [12, 13]. Its intended effect on the microenvironment was not noticeable in soil properties [14]. Rehabilitation of degraded steppes requires revegetation with an herbaceous layer, contributing to the improvement of soil quality [15]. Research on cactus pear concluded that it is a food of nutraceutical and functional importance [16, 17]. Cactus leaves are low in proteins and used as dairy cattle fodder [18]. It imparts better flavor and quality to the milk and enhances butter color [19]. In the cultivation of cactus pear, cladode and prickle pear fruits are produced. Cladodes ("nopal") with a production of 600,000 metric tones per annum in Mexico is consumed as a vegetable.

Cactus pear that has a high-water use efficiency (WUE) might play an increasingly important role in the agricultural systems of arid and semiarid regions under increased atmospheric CO_2 concentrations in the future [20]. For example, a doubling of atmospheric CO_2 concentration would lead to an increase in cactus pear biomass production from 23 to 55% due to increased WUE [4, 20]. Crassulacean acid metabolism (CAM) plants as prickly pear opens its stomata during the night to fix CO_2 as malic acid and converts it into sugar during the day [21]. Reduced rainfall caused by the climatic change will favor CAM plants with higher WUE [22], while the high potential productivity of *Opuntia* species may be used to slow the tendency of increasing atmospheric CO_2 levels [20].

1.2. Cactus pear production

About 100,000 ha are devoted to *Opuntia* fruit and cladode commercial production in more than 20 countries [23]. Commercial plantations use two types of production systems: traditional or intensive (microtunnel system). Prickle pear production with densities of 600–700 plants ha^{-1} allows easy machinery access, making orchard management more efficient [24]. As production declined by 29% between 2003 and 2010 [25], autosustainable farms of higher densities were required. New traditional systems plant 2–3 years old cladodes with a density of 15,000–40,000 plants ha^{-1}. Commercial plantations are located in regions with summer rainfall of 600–800 mm, providing yields from 30 to 80 tones ha^{-1} [26]. The *Opuntia* plant blooms once a year in Italy, while in countries like Chile and Israel, it blooms twice. Harvesting season differs according to variety, agro-climatic conditions, and forced blooming mechanisms [27].

The microtunnel system uses planting densities of 120,000–160,000 plants ha^{-1}, and beds with plastic sheet covers reduce the risk of frost damage in winter. Higher yields of 263 tones ha^{-1} are obtained during the entire year [9, 28]. Transparent plastic microtunnels produce cochineal, and the plants resisted three cycles, compared with green raffia canvas cultivars that resisted two cycles [29, 30].

Foreign markets, primarily the USA and Canada, are a growing opportunity for fresh and processed nopal edible cactus stems. The nutraceutical characteristics of nopal have raised interest in European and Asian markets [31]. Cladode stems are beneficial for the treatment of various diseases [32], being a good source of vitamin C [33], minerals [34], and soluble and insoluble fibers [31].

After conducting market studies to confirm that the future product will find markets, cactus pear-processing industries will be established [6, 35]. Cactus pear products may find a valuable niche in consumers that make healthy lifestyle and eating choices. It also includes vegetarians, health centers, and hospitals, where diet and medical care demand high-fiber, high-vitamin content, antioxidants, sugar-free foods, and similar attributes. It is critical to ascertain whether any competitors exist for the product and how successful they are.

1.3. Cactus pear composition and chemical contents

The cactus-pear fruit is an oval, elongated berry, with a thick pericarp, and a fleshy and juicy pulp that contains many hard-coated seeds [36]. The pericarp composed of a thick peel of commercially ripe fruits of *Opuntia ficus-indica* (L.) Mill. accounts for 33–55% of the fruit weight. Sepulveda and Saenz [37] cultivated this variety in Chile, and the proportion of peel with respect to the fruit weight was of 50.5 g (100 g^{-1}), with 49.5 g of edible fruit, 78.9% of pulp, and 20.1% of seeds. Research carried out with different varieties encountered that the pulp weight accounts from 45 to 67% of the total fruit weight, and the seeds volume within the pulp varied from 2 to 10%. The large variability depends on cultural practices, fruit load, climate, and harvesting season [38–40].

The pulp presents high pH values (5.3–7.1) and very low acidity (0.05–0.18%) that strongly influences the processing operations [41, 42]. No differences in soluble solids, acidity, and pH were noted in summer- and autumn-harvested fruits [43]. Sugars of the reducing type represent 10% of the pulp and range from 10–17°Brix. The fruit pulp is very sweet being glucose the predominant sugar and fructose the second sugar [37, 44, 45].

Cactus pear is higher in vitamin C than fruits such as apple, pear, grape, and banana [46]. *Opuntia ficus-indica* (L) Mill. shows a vitamin C content ranging from 180 to 300 mg kg^{-1} [47–49]. Ascorbic acid content was higher in summer-harvested fruits, while fruit weight was higher in autumn-harvested fruits [43]. Pectin and mucilage presence influences fruit pulp viscosity due to their capacity to hold and bind water. Total pectin content of cactus pear fluctuates from 5.32 to 14.19%, while the mucilage content varies between 3.78 and 8.5% [50].

Cactus pear pulp presents a good source of potassium (217 mg 100 g^{-1}) and a low level of sodium (0.6–1.19 mg 100 g^{-1}) being an advantage for people with renal and blood pressure problems [37]. Prickly pears are rich in calcium and phosphorus, 15.4–32.8 and 12.8–27.6 mg 100 g^{-1}, respectively [37, 51].

Cactus fruits are rich sources of yellow orange betaxanthins and red-violet betacyanins [32, 45, 52, 53]. Five different betaxanthins and six betacyanins structures have been identified in different *Opuntia*-colored fruits [32, 39, 54] being the ratio and concentration of betalains responsible for the yellow, orange, red, and purple colors. Cactus pear fruit have been suggested as a promising source of red and yellow food colorants for use at neutral pH [45], being used in yoghurts and ice creams [55]. Purple fruits are a source of betalains, a potential antioxidant and colorant similar to the pigment obtained from red beet that is widely used in the food industry [56].

1.4. Harvest timing and equipment

The cactus pear is a nonclimacteric fruit, and once harvested null ripening takes place. At 20°C, it has low ethylene production (0.2 nlg^{-1} h^{-1}), a low respiration rate (20 μLCO_2 g^{-1} h^{-1}), and it is not sensitive to ethylene [57]. It is important to collect the fruits at the optimal ripening stage for processing and consumption [36]. Fruit parameters such as size, peel color variation, firmness of the fruit, total suspended solids (TSS), and fall of the glochids determine optimal harvest period [6, 47]. Fruit harvest takes place after the soluble solids content exceeds 12°Brix.

When the peel coloration is halfway toward that of the fully ripened fruit, the TSS achieves 12–15%, depending on the cultivar. At this period, the fruit is at its best quality for consumption or for storage. The fully ripened fruit is not in the best condition for storage and is too soft for handling. It is best to begin harvesting early in the morning at the lowest possible air temperature to avoid glochids release. The fruits remain at low temperature, which reduces dehydration and infestation [6]. Fruit damage to the stem-end leads to attack by pathogens and fruit decay, so careful harvesting by twisting fruit from the stem or cutting fruit with a small piece of stem attached is required [58, 59]. However, during the postharvest period, the need to remove the attached pieces of cladode from the fruits will be required.

Fruits are collected in buckets or trays and left to cure in the sun, drying out any wounds and allowing the glochids to loosen. Fruit are laid-down in the field on beds of straw covered with plastic mesh. After removing the dry glochids with brooms, the fruit should be quickly packed and transported to a cool or refrigerated area. This is essential for long-term storage, avoiding dehydration and mold development on the fruit. Fruit packed in ventilated wooden or plastic crates, 4.5 kg cartons, or in single- or double-layer tray cartons [60] due to color, size, and quality condition [48]. Large fruits wrapped in tissue paper reduce scuffing and other physical injuries.

1.5. Postharvest of prickle pear fruit and shelf life

Cactus plantations in semiarid lands produce large quantities of cactus fruits with short shelf life where production rapidly surpasses demand. Improvements in shelf life will allow international refrigerated shipments of 3-weeks, withstanding 2°C fruit fly quarantine treatments and maintain its quality without the use of fungicides [61]. Extending shelf life opens up the possibility of supplying good-quality cactus pear fruit during periods of short supply in European markets [62].

Cactus pear fruits present a limited shelf life, even in cold storage, which has stimulated interest in obtaining processed items with increased shelf life [63]. High pH nonacid foods [64, 65] require of a sterilization treatment of 115.5°C or greater [66] to avoid pathogenic microorganisms growth such as *Fusarium* spp., *Alternaria* spp., and *Penicillium* spp. Stabilization procedures have detrimental effects on some sensory parameters of fresh fruits, such as color and flavor. Green cactus pear juice treatment with ultrasound technology reduced total cell counts and enter-bacteria to levels of no detection [67]. Juices presented good quality parameters, and ultrasound contributed to the release of bioactive compounds.

Storage of this perishable fruit can withstand from 2 to 5 weeks at temperatures ranging between 5 and 8°C with very humid environments (90–95% RH). Several factors can limit storage life such as decay, dehydration, and chilling injury [63, 68]. Chilling injury (CI) symptoms include pitting, surface bronzing, dark spots on the peel, and increased susceptibility to decay. Summer-harvested fruit was more chilling sensitive than autumn-harvested fruit [43].

Hot water treatments by immersion for a short time reduced microbial load without damaging the fruit [69]. In Italy, immersing the fruit for 5 min in hot water at 55°C extended the shelf life of cactus pear fruit cv. *Gialla* by 4–6 weeks without any prior chemical treatment [62]. Firmness and external appearance remained for 31 days with green and red prickle pears immersed during 2 min in water at 55°C [70]. After fruit hot water immersion at 52°C for 3 min, a microscope viewed prickle pear wounds and cracks sealing [71]. This left a relatively homogeneous surface from the fusion of the layers of wax, which provided real protection from fungi attack. Similar results appeared with hot air treatments at 37°C for 12 and 24 h [71].

Controlled atmosphere storage of minimally processed cactus pear fruits at 2°C in 10% CO_2 preserved them up to 20 days, maintaining high visual appearance, reduced tendency toward browning, and sugar content [72]. Passive-modified atmosphere packaging limited decay and increased the marketability of peeled cactus pear, as long as fruits were stored at low temperature [73, 74]. Ready-to-eat fruit storage in 5% O_2 and 30% CO_2 caused a selective suppression of the growth of different microbial populations [75]. In another experiment with ready-to-eat peeled fruits, unpackaged samples reported a higher mean pH value than modified atmosphere samples. However, samples packed in modified atmospheres reached the limit of marketability after 9 days [76, 77].

2. Harvesting devices

There are different ways of harvesting prickle pears, giving glochids the hardest time to workers [36, 78]. A simple method to pick the fruit with gloves is to twist it slightly and detach it (**Figure 1a**). The gloves covered with hair thorns end up in the worker's skin the next time it is used. Kitchen tongs (**Figure 1b**) detach prickle pears easily and keep the worker at a safe distance. After washing the tongs, the glochids disappear without ever touching worker's skin. Knives or scissors (**Figure 1c**) can remove the fruit from the cladode without damaging either the fruit or the cladode. Motor-blade harvesters (**Figure 4b**) provide a simple detaching method without entering in contact with glochids [79].

During fruit picking, workers cyclically perform many repetitive activities with the muscles of the hands and arms, causing excessive strains that increase occupational illnesses and decrease work productivity [60, 80]. A pruning shear developed in Egypt showed a 7.2% increase in labor productivity (kg day^{-1}) in comparison with handpicking. Microbiological activity was similar for handpicking and by using the pruning equipment [81].

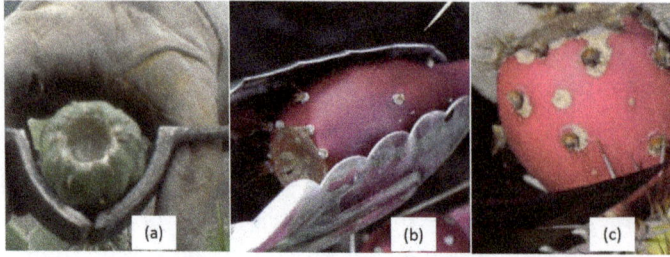

Figure 1. Harvesting methods using (a) gloves, (b) kitchen tongs, and (c) scissors.

2.1. Design of harvesting devices

The new harvesting tools should present high fruit detaching yield without producing hand pain. The light scissor device (**Figure 2a**) weighs only 412 g (**Table 1**), while the thimble cutter (**Figure 2b**) uses cast-iron covers to limit hand damage by thorns. The motorized equipment (**Figure 4b**) uses a light motor coupled to an abrasive disk being the distance between them adjusted for easy access to the prickle pear-cladode interface. At a speed of 1000 RPM, the device cuts the prickle pear in 1.1 s, consuming 1.36 A. A battery of 200 Ah would cut 147 prickle pears before discharging [79].

Ergonomic design criteria for pruning shears and harvesters should be included [81], considering a handle length longer than the widest part of hand (10–15 cm) and a strong spring, which stays set when pruning and reduces forceful exertions when opening. Another consideration includes grip span, wire cutting, and a rubber absorber to protect the wrist [81]. Grasping under internal forces, geometric and gravity sensors has been studied [82].

The third harvesting tool used a stainless steel wire to detach the prickle pear from the clad-ode (**Figure 3**). This light device contains a thimble for inserting the ring finger and another thimble-structure fixed to the glove within the palm of the hand. After moving the ring finger, both thimble-structures separate, stretching the wire and cutting the fruit that falls down into a storage pot. A spring between the thimbles (**Figure 3**) reduces the force exerted during the cut. A tactile sensor (mod HT201, TEKSCAN, USA) measured the force consisting

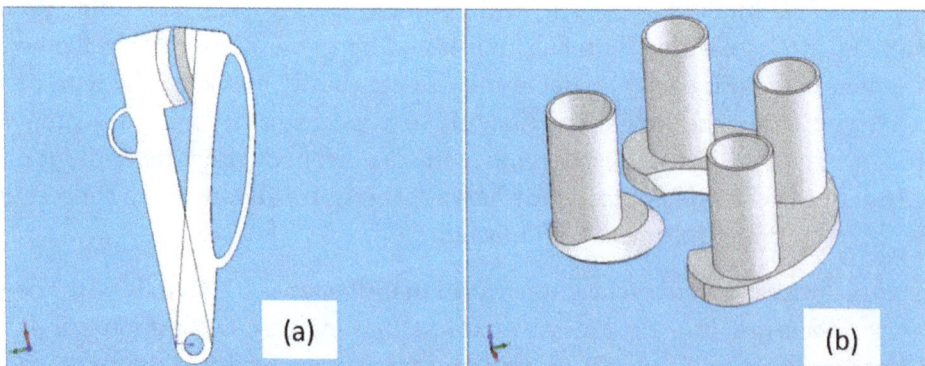

Figure 2. Proposed manual cutting (a) scissor tool and (b) thimble tool.

	Cutting tool				
	Disk	**Scissor**	**Thimble**	**Motor**	**Wire**
Weight, g	725	412	705	620	100
Diameter, mm	45	45	62	15	12
Length, mm	86	86	98	50	30
Pain	Small	Null	Regular	Vibration	Null
Yield, pears h^{-1}	300	275	350	300	350
Straight cut, %	41	48	39	35	90
Large hand worker with glove cutting efficiency, %	81	79	91	94	90
Large hand worker without glove cutting efficiency, %	83	81	92		88
Small hand worker with glove cutting efficiency, %	60	69	59	94	83
Small hand worker without glove cutting efficiency, %	63	71	61		81

Table 1. Performance of the different harvesting devices.

of a piezo-resistive material sandwiched between two pieces of flexible polyester and only about eight mils thick. The contact area is 9.53 mm and can measure up to 130 N. The force measured during cutting varied between 90 and 120 N according to the prickle pear maturity.

2.2. Evaluation of harvesting devices

Table 1 compares the yield, cutting quality, and pain generated to the workers for each cutting device. The effect of workers' hand size and glove use on cutting performance was studied.

Figure 3. Stainless steel harvester tool to detach fruits.

Figure 4. Device (a) slicing the pear straight, (b) electrically driven, (c) cutting with a simple knife, (d) slicing the fruit poorly.

Different workers performed daily during the autumn 2016 to obtain harvest data. **Table 1** compares the efficiency of the tools and although one device can provide a great yield if the cuts are not straight (**Figure 4d**), tool performance is poor. The stainless steel wire cutter cuts 350 pears h^{-1}, and 315 cuts were completely straight. Although the motorized device (**Figure 4b**) cuts 300 fruits in 1 h, it was harder to operate in cactus having pears lying close together (**Figure 4c**). As 35% of the fruit cuts were straight, only 105 were cauterized there. The prickle pear with the diagonal cut (**Figure 4d**) had to be sliced again, causing losses in the final harvest productivity. Productivity declined as day progressed for all harvesting devices.

Table 1 summarizes size (diameter and length) and weight of cast iron cutters. Reduction in weight to less than 750 g in the first three devices that resulted from 30% metal cover removal; the scissor device was the lightest. The thimble cutter was the shortest, but the worker's hand suffered pain, which remained for 2 months. In all cases, cutting efficiency decreased by using carnage gloves, which are clumsy and do not avoid the contact of glochids with the skin. The performance of the harvesting devices was compared (**Figure 4a**), showing the effect of the glove and the hand size. A reduction of 31% in fruits collected by small-hand workers that used gloves was noted when they employed the thimble-type tool, **Table 1**. Thimble-type tool size depends on worker's finger size; prickle pear size did not allow the cutter to operate easily.

The gripper hole of the disk-type prototype depends on worker's finger size; hand size affects cut quality and decreases productivity by 20%. The cutter moved by the thumb had no guide and was unable to enter between packed prickle pears on cactus plants (**Figure 4c**). Workers having small fingers produced many poor cuts, as they were unable to press hard with the thumb (**Figure 4d**). The prickle pear cannot be picked-up easily, but this light prototype does not cause pain to the worker.

The cutting wire of the wire-cutting device (**Figure 3**) is changed every week. However, it is adjustable to the hand size and does not leave pain to the worker. Its productivity is almost 6 pears min^{-1}, and it is impossible to cut the fruits diagonally.

In the future, intensive productions in microtunnels will be common. The space between rows is shorter, requiring robotic systems, similar to fruit-collecting robots within greenhouses [83–85]. A new mechanism of four degrees of freedom positions the cutting tool over cactus plants. The tractor provides support and power to the arm for mechanical harvesting of prickle pears [86]. At the time, no reports on cutting speed and efficiency are available.

3. Cauterizing equipment

Cauterization is the process of destroying tissue with electricity and is widely used in modern surgery. The cauterization of this equipment uses a thermal source to create heat transfer between the source and the fruit cells. Cells heal creating a diaphragm-type surface, which reduces fruit water loss. The two developed cauterization processes increased prickle pear shelf life without requiring hot water tanks for fruit immersion in the field. The system patented by Hahn in 2016 [87] cauterizes the prickle pear with a flat hot metallic surface. After removing the peduncle, an electric resistance contacts the fresh pulp. The trimmed peduncle ensures a better heat transfer between the fruit and the electric heating resistance of 200°C. The cells of the viscous pulp expand forming a diaphragm-type cover 7 h after cauterization that prevents dehydration [88]. The other equipment works with ice, and heat transfer occurs from fruit to ice.

3.1. Hot cauterizer

The resistance-cauterizing machine (**Figure 5a**) uses two pneumatic pistons and its operation is explained in several paper [89, 90]. At the end of the top piston, a Celeron™ machined cylinder is fixed and encloses a circular heating resistance with a flat contact surface. The equipment works following three cycles (**Figure 5a–c**), starting and finishing when both the pistons remain retracted. Once the pear is detected, the piston moves downwards to exert pressure over the fruit fixed within the cup. An ATM 89C51 microcontroller board (**Figure 5e**) adjusts the heating period responsible for sealing the pear. The microcontroller board controls the two $^3/_2$ way solenoid pneumatic valves (**Figure 5d**), which activate both pistons. A closed loop controller (mod E5CS-V, OMRON, Kyoto, Jap) maintains the heating resistance temperature constant by using a "J type" thermocouple. After 30 s, the embedded system interrupts the signal applied to the solenoid valve. The pneumatic pistons worked with a pressure varying between 100 and 200 kPa. A force of 20 N cm^{-2} supported by the piston applies 200 kPa of pressure to the pear; a pressure of 275 kPa destroyed immature cactus pears, separating the peel from the pulp.

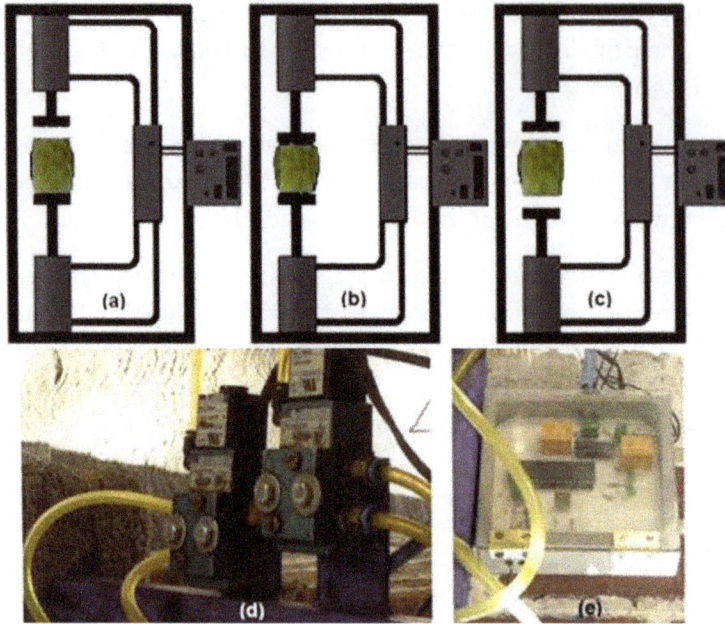

Figure 5. Heating cauterizer (a) lifts lower piston, (b) pressurizes the prickle pear, (c) finish treatment, (d) electro valves, and (e) microcontroller.

3.2. Experimental design and evaluation

The experimental design [89] consists of different treatments using four different pressures (50, 100, 150, and 200 kPa) and four different temperatures: 150, 180, 200, and 240°C. Each group consisted of 100 green sliced cactus pears numbered according to the applied pressure and temperature. After cauterization, the pears were stored at ambient temperature of 20 ± 3°C for 2 months. Visual observations rejected daily rotting fruit and shelf life for each group determined. Images of the cauterized area from five randomly selected pears were acquired with a digital camera (model WAT-250D, Watec, New York, USA) coupled to a stereoscopic microscope (model ZM160, ZEIGEN, D.F., Mexico) which magnified 40×. A nine W ring composed of 40 leds (model LED3000, LUXO CORP, New York, USA) provided 3600 foot-candles, which illuminated the sliced pear. Acquisition of images occurred after cauterization, 5 and 30 days after.

Measurements of pulp temperature during cauterization limited it to 70°C to avoid biochemical pulp changes [91]. A NTC thermistor (model STS-BTA, VERNIER, Beaverton, USA) with a resolution of 0.1°C and an accuracy of ±0.5°C was inserted in the fruit 0.5 cm beneath the cauterizing area. Pears were heated at 200°C using a pressure of 150 kPa, acquiring temperature measurements every second for a period of 45 s with a datalogger (VERNIER, Beaverton, USA).

Each pear was weighed every 5 days and the average water loss obtained. A machine texture analyzer (Zwicki Line Z0.5, ZWICK/ROELL, Germany) measured pulp firmness [90]. Measurements of firmness in the axial direction of the fruit used a 2.5 cm-diameter flat cylindrical probe. The force necessary to cause a deformation of 3 mm was determined to avoid fruit damage. Pulp firmness measurement happened 1 hour after cauterization, allowing time for the pulp to cool [90]. The value of pH was measured using a pH electrode and pH meter (HANNA, model HI

9021, USA), every 15 days after slicing ten prickle pears [92]. Every 15 days during the 60 days of storage, pulp firmness and pH of five fruits selected randomly were measured after slicing the pear just below the cauterized diaphragm.

3.3. Cold cauterizer

Freezing used for the long-term preservation, form ice crystals, and the final quality of the frozen product depends on the size and shape of these ice crystals [93]. Growth of ice crystals occurs after nucleation, after adding water molecules to the already formed nuclei. Slow freezing (0.02–0.2°C min^{-1}) enhances the formation of large extracellular ice crystals [93], altering the cellular membrane transport properties and consequently losing its ability to act as a semipermeable membrane; leeching of cellular substances and water loss is noted [94]. Growth of extracellular crystals caused tissue shrinkage and cell collapse [94].

The cold cauterizer is based on a pressure shift freezing system that will never arrive to zero degrees. Nucleation does not exists, as pressure release occurs very quickly [95]. Pressures of 200 and 150 kPa were applied to prickle pears, and the temperature within a pear was measured with a NTC thermistor inserted 0.5 cm from the sliced surface. Firmness, pH, and marketable fruit monitoring during storage provided results of the cauterization efficiency.

The device consists of a stainless steel vessel thermally insulated by a 5-cm thick glass fiber layer. A 3-cm thick ice bar introduced in the base of the equipment remains solid throughout 1 day. The fruit was manually placed within a hollow steel tube having a diameter of 7.5 cm so that two fingers can hold it (**Figure 6a**). A rack-and-pinion gear mechanism moved by a crank lowers the fruit held by a two finger-gripper (**Figure 6b**). These fingers have a spring that always press the fruit and do not drop it. Upon reaching the base, as the fruit touches the ice and cannot advance the fingers will open (**Figure 6c**).

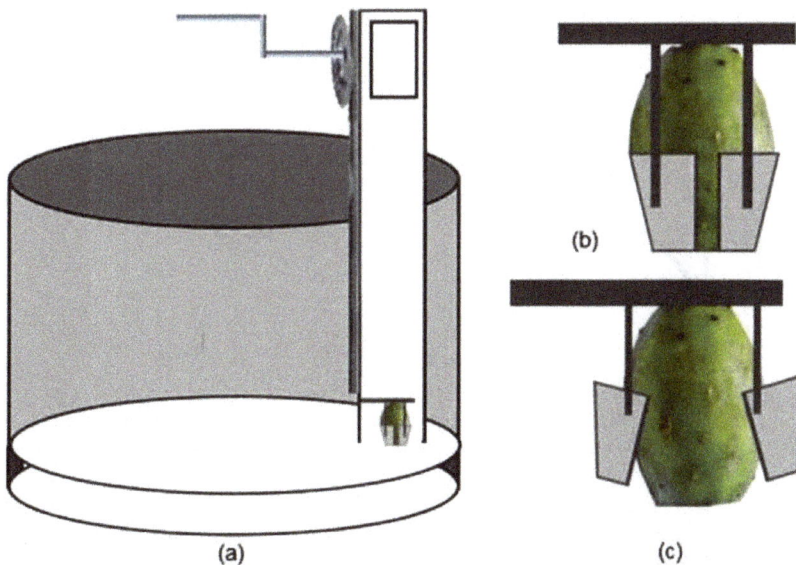

Figure 6. Cooling cauterizer (a) equipment, (b) during fruit transportation, and (c) during contact with ice.

3.4. Hot and cold cauterizer comparison

Surface pulp temperature increased to 56°C after applying a temperature of 200°C, while maintaining a pressure of 100 kPa (**Figure 7**). In the case of the cold cauterizer after 30 s, the temperature reached only 8°C, so no ice crystal building occurred. The temperature time constants were similar for hot and cold cauterization.

Cactus pears rotted after 15 days when not sliced and cauterized; slicing is an effective technique for getting a better contact area with the cauterizing surface. Fruits heated at 200°C using a pressure of 100 kPa began to rot after 37 days, but 50% of the fruit was still marketable after 63 days [90].

Quality parameters were measured on prickle pears during storage (**Table 2**) after hot cauterization (HT) treatment of 200°C@100 kPa and cold cauterization (CC) treatment of 100 kPa@0°C (ice). There were no significant differences encountered by ANOVA for both cauterizer treatments for fruit water loss during the first 45 days of storage. Hot and cold cauterizing processes presented average water loss after 2 months of storage of 0.9 and 4.5%, respectively. Significant differences appear after 60 storage days for hot treatments that applied pressures of 200 kPa [90]. The effect on Brix and pH on stored fruits after hot and cold cauterization was not significant, as shown by the ANOVA analysis. After 60 storage days, cold-cauterized pears increased its average soluble total solids by 7% over hot-processed fruits because of ripening, and decreased 8% its firmness (**Table 2**). Firmness showed significant differences in the cold treatment after 45 of storage as longer fruits began to decay and mature. As the fruit gets soft with a firmness of 13 N cm^{-2}, it gets sweeter, giving values over 14.5°Brix. These firmness and Brix values occurred 45 days after CC processing, and 2 months after HC cauterizing. Cold cauterization processing extended fruit shelf life, but poor appearance was observed in 45% of the CC fruits after 2 months of storage.

Microscopic images show a great difference between HC and CC treatments being thickness of the healed tissue variable. **Figure 8d** shows the diaphragm formed as seal after 2 months of HC

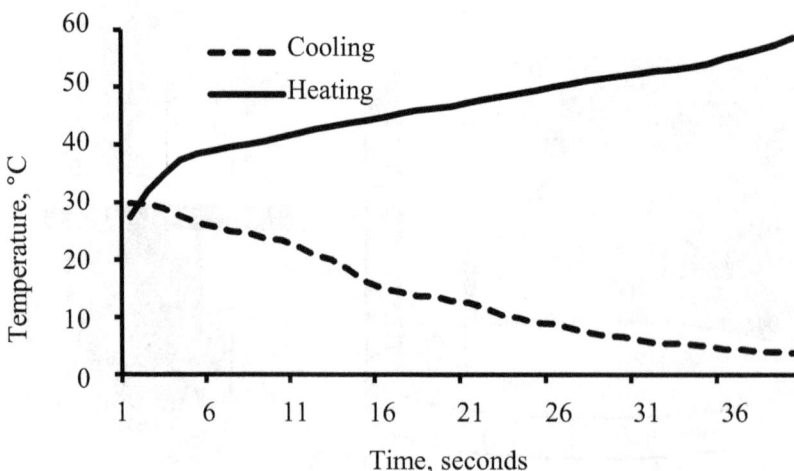

Figure 7. Effect of temperature inside the pear during hot and cold cauterization.

Variables	Cauterizer type	Measurement on:				
		Day 0	Day 15	Day 30	Day 45	Day 60
Water loss, %	H	0[a]	0[a]	0[a]	0.5[a]	0.9[b]
	C	0[a]	1.3[a]	2.1[a]	2.9[a]	4.5[b]
Firmness	H	16.4[a]	16.2[a]	15.7[a]	14.8[a]	13.1[b]
N cm^{-2}	C	16.38[a]	15.6[a]	14.5[a]	13.1[b]	12.1[b]
°Brix	H	13.05[a]	13.1[a]	13.15[a]	13.5[a]	14.19[a]
	C	13.1[a]	13.4[a]	14[a]	14.8[a]	15.1[a]
pH	H	5.32[a]	5.32[a]	5.39[a]	5.66[a]	5.79[a]
	C	5.32[a]	5.35[a]	5.41[a]	5.72[a]	5.88[a]
Fruits	H	0[a]	0[a]	0[a]	4[a]	15[b]
Decay, %	C	0[a]	0[a]	2[a]	16[b]	45[c]
Marketable	H	100[a]	100[a]	100[a]	98[a]	75[b]
Fruit, %	C	100[a]	100[a]	96[a]	82[b]	50[c]

[a-c]Values sharing letters in the same row are not statistically significantly different from each other (based on least significant differences (lsd) from ANOVA, calculated standard error of 0.08).

Table 2. Compared variation of prickle pear stored after hot and cold cauterization every 15 days.

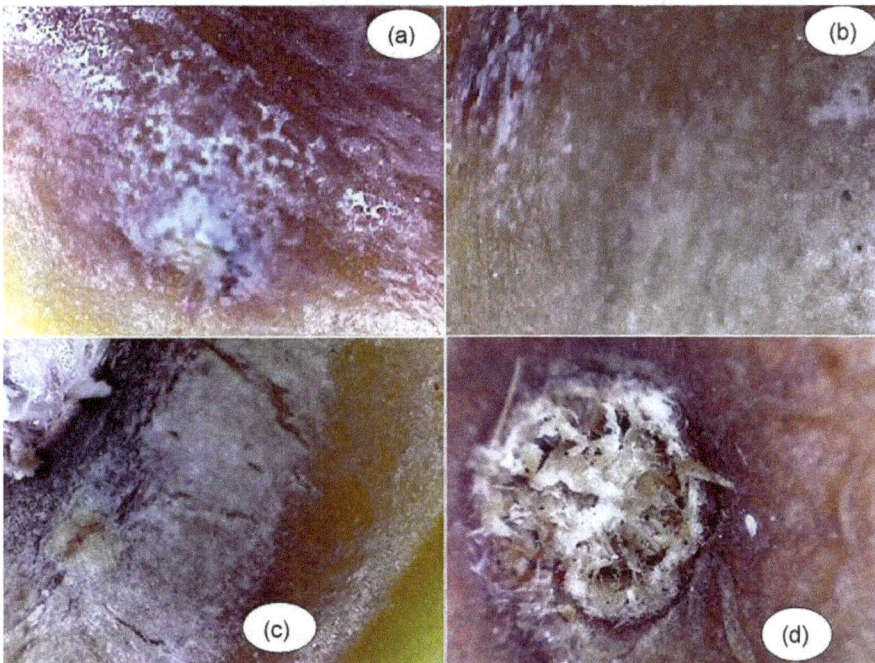

Figure 8. Microscopic image of (a) CC after 1 day, (b) CC shrinkage of cells, (c) HC seal after 1 month, and (d) HC seal after 2 months.

cauterization. One month before (**Figure 8c**), the seal was perfectly formed and did not allow any water loss. In **Figure 8a**, no seal appears and water movement appears after 1 month; temperature should be lower so that a thicker seal can appear and the cool-cauterizer process can be quicker.

4. Hot cauterizer line

The amount of prickle pears processed per hour is dependent on the time required for transferring heat from the resistance to the fruit in order to cauterize it. With a cauterization period of 30 s, only 120 fruits are cauterized per hour [90]. To increase the processing quantity, a packaging line was designed and implemented consisting of three sections. The first section removes the thorns, the second cauterizes the fruits, and the third packs them. The amount of cauterized pears per hour depends on the amount of cauterizing units operating in parallel, so if the line has 10 cauterizers, it will process 1200 fruits h^{-1} (approximately 250 kg h^{-1}). Several of these lines operating in parallel will increase even more the number of fruits processed.

Marquez and collaborators [25] found that 40% from the total prickle pear economical income was used for removing prickly pear glochids and packing; wooden boxes accounted for 35% of the crop revenue. Elimination of glochids in small processing plants takes place by running the fruits over rollers covered with hairs or nylon bristle brushes [6]. The first stage of the packaging line removes the glochids from the singulated prickle pears that advance slowly in-between two nylon brushes rotating in opposite directions. The brushes placed on each side of the V-type conveyor make the pears spin around removing the glochids without damaging the fruit.

Each prickle pear advances alone through the line and an articulated gate stops it; gates quantity depends on the number of cauterizer units. The controlled gate splits a pear from its neighbor by 15 cm, letting the fruit beneath the cauterizing piston (**Figure 9a**). A vision system analyzes the quality from each prickle pear as they advance through the line. After acquiring the image, an edge detection algorithm eliminates the contours (**Figure 10a**). The filtered image uses a mask to eliminate the glochids areas marked with green circles, making the rot in red detectable (**Figure 10b**). Once the high-quality fruit passes the vision system analysis, a signal turns on a cutter for removing the peduncle from the prickle pear. The fruit turned with the bare pulp upward, waits for the contact of the hot plate (**Figure 9b**). Each piston works independently since the displacement of each piston arm depends on fruit size. The process ends after 30 s of cauterization by lifting the top pistons (**Figure 9c**).

The third section of the line is the one that packs the fruits. Prickle pears move carefully through the conveyor belt, as fruits have no protection at the top side, after peduncle removal and fruit cauterization. Photodetector A (**Figure 11**) detects fruit movement over the conveyor. Once the conveyor rotates over the sprocket, the pear will fall unless a hard-plastic structure avoids it; the frame rotates the fruit by 90° maintaining the cauterized part looking up. Fruit turns around its center of gravity as shown in **Figure 11**, being the pear pulled downward by the band movement. The center of gravity causes fruit turning (**Figure 11**), and band movement pulls the pear downward.

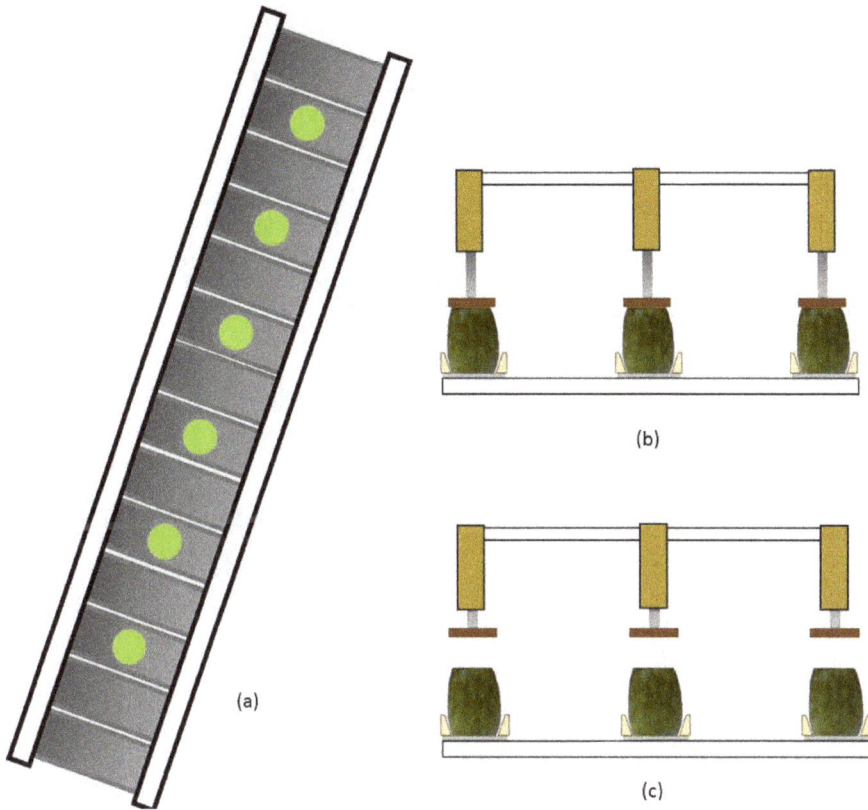

Figure 9. Packing line showing the (a) conveyor line, (b) cauterization units, and (c) end of process.

A camera or a capacitive sensor (B) transmits a signal to a movable gate to close. The gate stops the fruit falling by gravity, ensuring that the arm that is going to transport it to the packing boxes is in its right position. When sensor C perceives the pear standing over the gateway and sensor D detects the proper position of the gripper, the movable gate opens and the fruit will fall into the gripper fingers. The fingers move the fruit to the packing box.

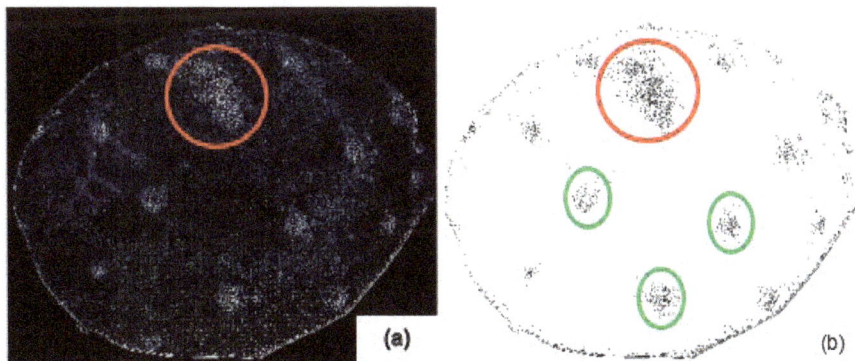

Figure 10. Image of a rot fruit after (a) edge detection and (b) filtering and noise removal.

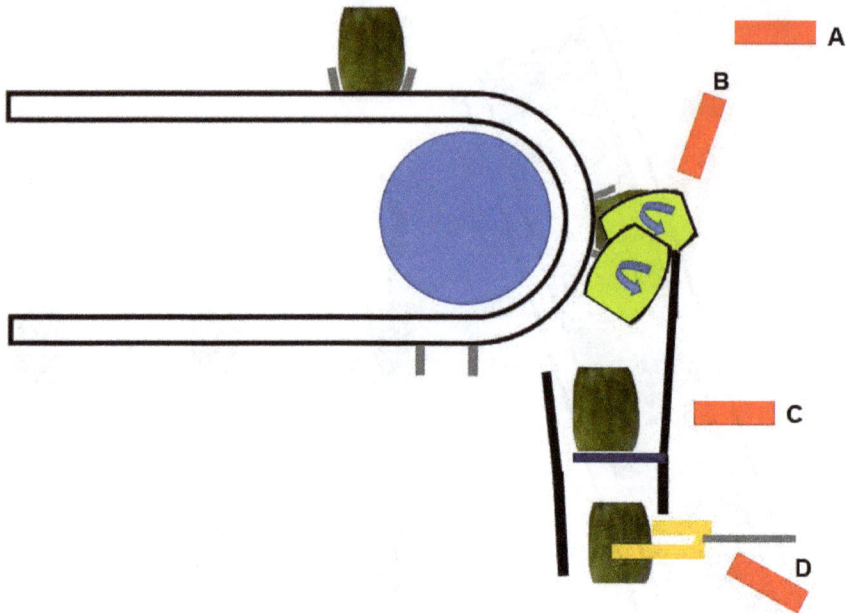

Figure 11. Packing mechanism for prickle pears.

5. Conclusion

An increase of the shelf life of prickle pear fruits can make them marketable in developed countries, considering the fruit nutraceutical and chemical functional properties. The first step to increase shelf life is to harvest fruits carefully as mechanical damage to the stem-end leads to attack by pathogens and fruit decay.

There are different ways of harvesting prickle pears, which include gloves, kitchen tongs, motor-blade tools, knives, and scissors. An important consideration when designing a tool for prickle pear fruit detaching is to avoid skin contact with glochids. Weight and another physical dimension of the tools, productivity, and quality of cuts were compared during harvesting trials using five tools. The stainless steel wire tool was the lighter and the more efficient; workers did not suffer any pain after using it. Ninety percent of the collected fruits had straight cuts, and 350 fruits were detached hourly. The motorized system was difficult to operate between fruits close together over the cladode.

Cauterization increased prickle pear fruits' shelf life over 2 months. Hot and cold cauterizer equipment extended shelf life without pathogen damage as the treatment seals the fruit and avoids dehydration. The hot cauterizer places a hot surface at 200°C in contact with the fresh prickle pear pulp. The trimmed peduncle allows a better heat transfer, enhanced further by a pressurized contact of 100 kPa. The cold cauterizer replaced the hot surface by ice to process the fruit. Pulp firmness, pH, water loss, Brix, and percentage of marketable fruit were compared using both cauterizers. The hot cauterizer working at 200°C and 100 kPa extended shelf life, being 85% of the fruit marketable after 2 months. Fruits were stored at ambient conditions. An automated cauterizer line was introduced for processing one quarter ton every hour.

Author details

Tebien Federico Hahn

*Address all correspondence to: fhahn@correo.chapingo.mx

Chapingo Autonomous University, Texcoco, Mexico

References

[1] Griffith P. The origins of an important cactus crop, *Opuntia ficus-indica* (Cactaceae): New molecular evidence. American Journal of Botany. 2004;**91**:1915-1921

[2] De la Rosa HJ, Santana AD. El nopal: Usos, manejo agronómico y costos de producción en México. México: Editorial Conaza-Uach-Ciestaam; 1998. pp. 30-31

[3] Granados SD. El nopal: Historia, fisiología, genética e importancia frutícola. México: Editorial Trillas; 2003. pp. 5-10

[4] Barbera G. History, economic and agro-ecological importance. In: Barbera G, Inglese P, Pimienta E, editors, Agro-ecology, Cultivation and Uses of Cactus Pear. Rome: FAO; 1995. pp. 1-8

[5] Barbera G. Historia e importancia económica y agroecológica. In: Barbera G, Inglese P, Pimienta E, editors. Agroecología, cultivo y usos del nopal. Rome: FAO; 1999. pp. 1-12

[6] Saenz C, Berger H, Rodríguez-Félix A, Galletti L, Corrales García J, Sepúlveda E, Varnero MT, García de Cortázar V, Cuevas García R, Arias E, Mondragón C, Higuera I, Rosell C. Agro-industrial utilization of cactus pear. Rome: FAO; 2013. E-ISBN 978-92-5-107987-4

[7] Selmi S, Khalfaoui A, Chouki S. Cactus in Zelfene (Tunisia): An alternative for rural development. Cactusnet Newsletter. 2002;**6**:5-9

[8] Gibson A, Nobel PS the Cactus Primer. Cambridge, Massachusetts. USA: Harvard University Press; 1986. 296 pp

[9] Flores-Valdez CA. Producción, industrialización y comercialización de nopalitos. In: Barbera G, Inglese P, Pimienta E, editors. Agroecología, cultivo y usos del nopal. Rome: FAO; 1999. pp. 97-105

[10] Potgieter JP. The influence of environmental factors on spineless cactus pear (*Opuntia* spp.) fruit yield in Limpopo Province, South Africa [thesis]. Bloemfontein: University of the Free State; 2007

[11] Panda SK, Behera SK, Witness Qaku X, Sekar S, Ndinteh DT, Nanjundaswamy HM, Ray RC, Kayitesi E. Quality enhancement of prickly pears (*Opuntia* sp.) juice through probiotic fermentation using Lactobacillus fermentum – ATCC 9338. Food Science and Technology. 2017;**75**:453-459. DOI: 10.1016/j.lwt.2016.09.026#doilink

[12] Neffar S, Beddiar A, Redjel N, Boulkheloua J. Effets de l'âge des plantations de figuier de Barbarie (*Opuntia ficus-indica* f. inermis) sur les propriétés du sol et la végétation à Tébessa (zone semi-aride de l'est algérien). Ecologia Mediterranea. 2011;**37**:5-15

[13] Snyman H. A greenhouse study of root dynamics of cactus pears, *Opuntia ficus-indica* and *O. robusta*. Journal of Arid Environments. 2006;**65**:529-542

[14] Su Y, Zhao L. Soil properties and plant species in an age sequence of *Caragana* microphylla plantations in the Horqin Sandy land, North China. Ecological Engineering. 2003;**20**:223-235

[15] Li X, Jia, X, Dong J. Influence of desertification on vegetation pattern variation in the cold semi-arid grasslands of Qinghai-Tibet Plateau, North-west China. Journal of Arid Environments. 2006;**64**:505-522

[16] Feugang JM, Konarski P, Zou D, Stintzing FC, Zou C. Nutritional and medicinal use of Cactus pear (*Opuntia* spp.) cladodes and fruits. Frontiers in Bioscience. 2006;**11**:2574-2589

[17] Piga A. Cactus pear: A fruit of nutraceutical and functional importance. Journal of the Professional Association for Cactus Development. 2004;**6**:9-22

[18] Brasil JN, Jereissati ES, Santos MRA, Campos FAP. In vitro micropropagation of *Nopalea cochenillifera* (Cactaceae). Journal of Applied Botany and Food Quality. 2005;**79**(3):160-162

[19] Oliveira VSD, Ferreira MDA, Guim A, Modesto EC, Arnaud BL, Silva FMD. Effects of replacing corn and tifton hay with forage cactus on milk production and composition of lactating dairy cows. Revista Brasileira de Zootecnia. 2007;**36**(4):928-935

[20] Nobel PS. Environmental biology. In: Barbera G, Inglese P, Pimienta E, editors. Agro-ecology, Cultivation and Uses of Cactus Pear. FAO Plant Production and Protection Paper No. 132. Rome: FAO; 1995. pp. 36-48

[21] Pimienta-Barrios E, Zannudo J, Yepez E, Novell PS. Seasonal variation of net CO_2 uptake for cactus pear *(Opuntia ficus-indica)* and pitayo *(Stenocereus queretaroensis)* in a semi-arid environment. Journal of Arid Environments. 2000;**44**:73-83

[22] Le Houérou HN. Drought-Tolerant and Water-Efficient Fodder Shrubs (DTFS), Their Role as a "Drought Insurance" in the Agricultural Development of Arid and Semi-Arid Zones in Southern Africa; 1994. WRC Report No. KV 65/94. 139 p

[23] Nobel PS. Environmental Biology of Agaves and Cacti. Cambridge, New York; Cambridge university press, UK. 1988

[24] Corrales GJ, Flores VC. Nopalitos y Tunas: Producción, Comercialización, Postcosecha e Industrialización. Primera edición. México: Universidad Autónoma Chapingo; 2003

[25] Marquez SR, Torcuato C, Almaguer G, Colinas MT, Gardezi AK. El sistema productivo del nopal tunero (*Opuntia albicarpa y O. megacantha*) en Axapusco, Estado de México: problemática y alternativas. Revista Chapingo Serie Horticultura. 2012;**18**(1):81-93

[26] Pimienta-Barrios E, Barbera G, Inglese P. Cactus pear (*Opuntia* spp. Cactaceae) International Network: An effort for productivity and environmental conservation for arid and semiarid lands. Cactus and Succulent Journal. 1993;**65**:225-229

[27] Inglese P. Plantación y manejo de huertos. In: Barbera G, Inglese P, Pimienta E, editors. Agroecología, cultivo y usos del nopal. Rome: FAO; 1999. pp. 82-96

[28] Blanco-Macías F, Valdez-Cepeda R, Ruiz-Garduño RR. Intensive production of cactus pear under plastic tunnels. In: Nefzaoui A & Inglese P, editors. Proceedings 4th International Congress on Cactus Pear and Cochineal, 22-28 October, Hammamet, Tunisia. Acta Horticulturae. 2002;**581**:279-282

[29] Aldama-Aguilera C, Llanderal-Cázares C, Soto-Hernández M, Castillo-Márquez LE. Producción de grana-cochinilla (Dactylopius coccus Costa) en plantas de nopal a la intemperie y en microtúneles. Agrociencia. 2005;**39**:161-171

[30] Campos-Figueroa M, Llanderal-Cázares C. Producción de grana cochinilla Dactylopius coccus (Homoptera: Dactylopidae) en invernadero. Agrociencia. 2003;**37**:149-155

[31] Peña-Valdivia CB, Trejo LC, Arroyo-Peña A, Sánchez U, Balois R. Diversity of unavailable polysaccharides and dietary fiber in domesticated nopalito and cactus pear fruit (*Opuntia* spp.). Chemistry and Biodiversity. 2012;**9**:1599-1610

[32] Stintzing FC, Carle R. Cactus stems: A review on their chemistry, technology, and uses. Molecular Nutrition & Food Research. 2005;**49**(2):175-194. DOI: 10.1002/mnfr.200400071

[33] Betancourt-Domínguez MA, Hernández-Pérez T, García-Saucedo P, Cruz-Hernández A, Paredes-López O. Physico-chemical changes in cladodes (nopalitos) from cultivated and wild cacti (*Opuntia* spp.). Plant Foods for Human Nutrition. 2006;**61**:115-119. DOI: 10.1007/s11130-006-0008-6

[34] Rodriguez-Felix A, Cantwell M. Developmental changes in composition and quality of prickly pear cactus cladodes (nopalitos). Plant Foods for Human Nutrition. 1988;**38**:83-93

[35] Cuevas R, Masera O. Díaz R. Calidad y competitividad en la industria alimentaria rural latinoamericana a través del uso eficiente y sostenible de energía. In: Cuevas R, Masera O, Díaz R, editors. Calidad y competitividad de la agroindustria rural de América Latina y el Caribe. Rome: FAO; 2004. pp. 7-13

[36] Reyes-Agüero JA, Aguirre-Rivera JR, Hernández, HM. Notas sistemáticas y una descripción detallada de *Opuntia ficus-indica* (L.) MILL. (CACTACEAE). Agrociencia. 2005;**39**:395-408

[37] Sepúlveda E, Sáenz C. Chemical and physical characteristics of prickly pear (*Opuntia ficusindica*). Revista de Agroquimica y Tecnologia de Alimentos. 1990;**30**:551-555

[38] Barbera G, Inglese P, La Mantia T. Seed content and fruit characteristics in cactus pear (*Opuntia ficus-indica* Mill.). Scientia Horticulturae.1994;**58**:161-165

[39] Inglese P, Barbera G, La Mantia T, Portolano S. Crop production and ultimate size of cactus pear fruit following fruit thinning. HortScience. 1995;**30**:227-230

[40] Mondragon JC, Gonzalez SP. Native cultivars of cactus pear in Mexico. In: Janick J, editor. Progress in New Crops. Arlington: Va. ASHS Press; 1996. pp. 446-450

[41] Sepúlveda E. Cactus pear fruit potential for industrialization. In: Saenz C, editor. Proceedings of the International Symposium: Cactus pear and nopalitos processing and uses. Universidad de Chile, Santiago, and International Cooperation and Network on Cactus pear, Santiago, Chile; 1998. pp. 17-21

[42] Sáenz C. Processing technologies: An alternative for cactus pear (*Opuntia* spp.) fruits and cladodes. Journal of Arid Environments. 2000;**46**:209-225

[43] Schirra M, Inglese P, La Mantia T. Quality of cactus pear (*Opuntia ficus-indica* (L.) Mill.) fruit in relation to ripening time, $CaCl_2$ preharvest spray and storage conditions. Scientia Horticulture. 1999;**81**:425-436

[44] Russel CH, Felker P. The prickly pears (*Opuntia* spp. *Cactaceae*). A source of human and animal food in semiarid regions. Economic Botany. 1987;**41**:433-445

[45] Stintzing FC, Schieber A, Carle RC. Evaluation of colour properties and chemical quality parameters of cactus juices. European Food Research Technology. 2003;**216**:303-311

[46] Cheftel JC, Cheftel H. Frutta e Verdura. In: Cheftel JC, Cheftel H, editors. Biochimica e tecnologia degli alimenti. Bologna, Italy: Edagricole; 1983. pp. 147-240

[47] Cantwell M. Postharvest management of fruits and vegetable stems. In: Barbera G, Inglese P, Pimienta E, editors. Agro-ecology, Cultivation and Uses of Cactus Pear. Rome: FAO; 1995. pp. 120-136

[48] Piga A, D'Aquino S, Agabbio M, Schirra M. Storage life and quality attributes of cactus pears cv "Gialla" as affected by packaging. Agricoltura Mediterranea. 1996;**126**:423-427

[49] Sáenz C. Food manufacture and by-products. In: Barbera G, Inglese P, Pimienta E, editors. Agro-ecology, Cultivation and Uses of Cactus Pear. Rome: FAO; 1995. pp. 137-143

[50] Pena VBC, Sanchez UBA. Nopalito and cactus pear polysaccharides: Mucilage and pectin. Acta Horticulturae. 2006;**728**:241-247

[51] Sawaya WN, Khatchadourian HA, Safi WM, Al-Hammad HM. Chemical characterization of prickly pear pulp, *Opuntia ficus-indica*, and the manufacturing of prickly pear jam. Journal of Food Technology. 1983;**18**:183-193. DOI: 10.1111/j.1365-2621.1983.tb00259.x

[52] Castellar R, Obon JM, Alacid M, Fernandez-Lopez JA. Color properties and stability of betacyanins from *Opuntia* fruits. Journal of Agricultural and Food Chemistry. 2003;**51**:2772-2776

[53] Odoux E. Dominguez-López A. Le figuier de Barbarie: Une sorce industrialle de bétalaines? Fruits. 1996;**51**:61-78

[54] Felker P, Inglese P. Short term and long-term research needs for *Opuntia ficus-indica* (L.) Mill. Utilization in arid areas. Journal of the Professional Association for Cactus Development. 2003;**5**:131-151

[55] Stintzing FC, Carle R. Cactus fruits – more than just colour. Flussiges Obst. 2006;**73**(8):430-434

[56] Saenz C. *Opuntia* spp. Bioactive compounds in foods: A plus for health. Acta Horticulturae. 2006;**728**:231-239

[57] Cantwell M. Manejo postcosecha de tunas y nopalitos. In: Barbera G, Inglese P, Pimienta E, editors. Agroecología, cultivo y usos del nopal. Rome: FAO; 1999. pp. 126-143

[58] Ochoa J, Degano C, Ayrault G, Alonso ME. Evaluation of postharvest behavior in cactus pear (*Opuntia ficus-indica* L. Mill.). Acta Horticulturae. 1997;**438**:115-121

[59] Martínez-Soto G, Fernández-Montes MR. Cabrera-Sixto JM. Evaluación de la vida de anaquel de tuna (Opuntia ficus-indica). In Memoria VIII Congreso Nacional y VI Internacional sobre Conocimiento y Aprovechamiento del Nopal. Universidad Autónoma de San Luís Potosí. San Luis Potosí, SLP, México; 1999. pp. 26-27

[60] Kaur K, Dhillon WS, Mahajan BVC. Effect of different packaging materials and storage intervals on physical and biochemical characteristics of pear. Journal of Food Science and Technology. 2013;**50**(1):147-152. DOI: 10.1007/s13197-011-0317-0

[61] Rodriguez S, Casoliba RM, Questa AG, Felker P. Hot water treatment to reduce chilling injury and fungal development and improve visual quality of *Opuntia ficus-indica* fruit clones. Journal of Arid Environments. 2005;**63**(2):366-378. DOI: 10.1016/j.jaridenv.2005.03.020

[62] Schirra M, Agabbio M, D'Aquino S, Mc Collum TJ. Postharvest heat conditioning effects on early ripening *Gialla* cactus pear fruit. HortScience. 1997;**32**(4):702-704

[63] Ochoa-Velasco CE, Guerrero-Beltrán JA. Efecto de la temperatura de almacenamiento sobre las características de calidad de tuna blanca villanueva (*Opuntia albicarpa*). Revista Iberoamericana de Tecnología Postcosecha. 2013;**14**(2):149-161

[64] Sáenz C. Foods products from cladodes and cactus pear. Journal of the Professional Association for Cactus Development. 1996;**1**:89-97

[65] Piga A, Hallewin GD, Aquino SD, Agabbio M. Influence of film wrapping and UV irradiation on cactus pear quality after storage. Packaging Technology and Science. 1997;**10**:59-68

[66] Sáenz C, Cactus pear juices. Journals of the Professional Association for Cactus Development. 2001;**4**:3-10

[67] Cansino NC, Carrera GP, Rojas QZ, Olivares LD, García EA, Moreno ER. Ultrasound processing on green cactus pear (*Opuntia ficus-indica*) juice: Physical, microbiological and antioxidant properties. Journal of Food Processing and Technology. 2013;**4**:267. DOI: 10.4172/2157-7110.1000267

[68] Yahia EM, Saenz C. Cactus pear fruit. In: Yahia EM, editor. Postharvest Biology and Technology of Tropical and Sub-Tropical Fruits. Woodhead Publishing Co, England; 2010

[69] Shewfelt RL. Postharvest treatment for extending shelf life of fruits and vegetables. Food Technology. 1986;**40**(5):70-78. 89

[70] Berger H, Mitrovic A, Galletti L, Oyarzún J. Effect of hot water and wax applications on storage life of cactus pear (*Opuntia ficus-indica* (L) Mill.) fruits. Acta Horticulturae. 2002;**581**:211-220

[71] D'hallewin G, Schirra M, Manueddu E. Effect of heat on epicuticular wax of cactus pear fruit. Tropical Science. 1999;**39**:244-247

[72] Anorve MJ, Aquino BEN, Mercado SE. Effect of controlled atmosphere on the preservation of cactus pears. Acta Horticulturae. 2006;**728**:211-216

[73] Piga A, Del Caro A, Pinna I, Agabbio M. Changes in ascorbic acid, polyphenol content and antioxidant activity in minimally processed cactus pear fruits. LWT – Food Science and Technology. 2003;**36**:257-262

[74] Brito Primo DM, Martins LP, Lima AB, Silva SM, Barbosa JA. Postharvest quality of cactus pear fruits under modified atmosphere and refrigeration. Acta Horticulturae. 2009;**811**:167-172

[75] Corbo MR, Altieri C, D'Amato D, Campaniello D, Del Nobile M, Sinigaglia M. Effect of temperature on shelf life and microbial population of lightly processed cactus pear fruit. Postharvest Biology and Technology. 2004; **31**(1):94-104

[76] Cefola M, Renna M, Pace B. Marketability of ready-to-eat cactus pear as affected by temperature and modified atmosphere. Journal of Food Science and Technology. 2011;**47**(1). DOI: 10.1007/s13197-011-0470-5

[77] Cefola M. Renna M, Pace B. Marketability of ready-to-eat cactus pear as affected by temperature and modified atmosphere. Journal of Food Science and Technology 2014;**51**:25-33

[78] Lloret-Salamanca A, Lloret PG, Blasco M, Angulo MJ. Comparative morphological analysis of areole and glochids of two *Opuntia* species. Acta Horticulturae. 2015;**1067**:59-65. DOI: 10.17660/ActaHortic.2015.1067.7

[79] Hahn F. Photovoltaic prickle pear-harvesting tool. Natural Resources. 2013;**4**:263-265

[80] Callea P, Zimbalatti G, Quendler E, Nimmerichter A, Bachl N, Bernardi B, Smorto D, Benalia S. Occupational illnesses related to physical strains in apple harvesting. Annals of Agricultural and Environmental Medicine. 2014;**21**(2):407-411. DOI: 10.5604/1232-1966.1108614

[81] El–Ghany AMA, Horia MA, Nahed K, El-Bialee NM. Developed hand tool to pick prickly pear fruit. International Journal of Advanced Research. 2015;**3**(5):1307-1315

[82] Svinin MM, Kaneko M, Tsuji T. Internal forces and stability in multifinger grasps. Control Engineering Practice. 1999;**7**:413-422

[83] van Henten EJ, Hemming J, van Tuijl BAJ, Kornet JG, Meuleman J, Bontsema J, van Os EA. An autonomous robot for harvesting cucumbers in greenhouses. Autonomous Robots. 2002;**13**:241-258

[84] Martinez JL, Mandow A, Morales J, Pedraza S, Garcia-Cerezo A. Approximating kinematics for tracked mobile robots. The International Journal of Robotics Research. 2005;**24**(10):867-878

[85] Singh S, Burks TF, Lee WS. Autonomous Robotic Vehicle for Greenhouse Spraying. ASAE Paper No. 043091. St. Joseph, Mich.: ASAE; 2004

[86] Durán HM, García OG, Delgado JLMP, Kipping ER, Delgado CIJ. Structural design of a mechanical arm for harvest of cactus pear type Alfajayucan. Journal of Applied Research and Technology. 2016;**14**:140-147

[87] Hahn F. Equipo industrial para la cauterización de tunas. Patente IMPI, MX 343799 B Mexico DF. Concedida el 16 Nov 2016

[88] Hahn F. Cactus pear cauterizer increases shelf life without cooling processes. Computers and Electronics in Agriculture. 2009;**65**:1-6

[89] Hahn F, Salvador N, Cruz J. Thermal evaluation of cauterized prickle pears by an industrial machine. In: Proceedings of the CIGR-Agricultural Engineering Conference. Valencia. Asabe Technical Library, Michigan USA, Spain, 8-12 July 2012; 2012

[90] Hahn F, Cruz J, Barrientos A, Perez R, Valle S. Optimal pressure and temperature parameters for prickly pear cauterization and infrared imaging detection for proper sealing. Journal of Food Engineering. 2016;**191**:131-138. DOI: 10.1016/j.jfoodeng.2016.07.013.%20 0260-8774

[91] Merin U, Gagel S, Popel G, Bernstein S. Thermal degradation kinetics of prickly pear fruit red pigment. Journal of Food Science. 1987;**52**:485-486

[92] AOAC. In: Helrich K, editor. Official Methods of Analysis of AOAC: Food Composition; Additives; Natural Contaminants. Vol. II. Arlington: AOAC; 1990

[93] Fernández L, Otero B, Guignon PD. High-pressure shift freezing versus high-pressure assisted freezing: Effects on the microstructure of a food model. Food Hydrocolloids, Wrexham. 2006;**20**:510-522

[94] Delgado AE, Rubiolo AC. Microstructural changes in strawberry after freezing and thawing processes. Lebensmittel Wissenschaft und-Technologie. 2005;**38**:135-142

[95] Pham QT. Advances in food freezing/thawing/freeze concentration modelling and techniques. Japan Journal of Food Engineering. 2008;**9**(1):21-32

2

Optical Methods for Firmness Assessment of Fresh Produce

Jason Sun, Rainer Künnemeyer and
Andrew McGlone

Additional information is available at the end of the chapter

Abstract

This chapter is devoted to a review of optical techniques to measure the firmness of fresh produce. Emphasis is placed on the techniques that have a potential for online high-speed grading. Near-infrared spectroscopy (NIRS) and spatially resolved reflectance spectroscopy (SRRS) are discussed in detail because of their advantages for online applications. For both techniques, this chapter reviews the fundamental principles as well as the measured performances for measuring the firmness of fresh produce, particularly fruit. For both techniques, there have been studies that show correlations with penetrometer firmness as high as $r = 0.8 - 0.9$. However, most studies appear to involve bespoke laboratory instruments measuring single produce types under static conditions. Therefore, accurate performance comparison of the two techniques is very difficult. We suggest more studies are now required on a wider variety of produce and particularly comparative studies between the NIRS and SRRS systems on the same samples. Further instrument developments are also likely to be required for the SRRS systems, especially with an online measurement where fruit speed and orientation are likely to be issues, before the technique can be considered advantageous compared to the commonly used NIRS systems.

Keywords: produce, firmness, spatially resolved reflectance spectroscopy, near-infrared spectroscopy, optical methods

1. Introduction

Firmness is a major quality parameter in grading fresh produce, governed by the mechanical and structural properties of fruit. For producers, it can indicate ripeness and/or storage potential, and for consumers, it directly influences consumer acceptance and satisfaction. The industry standard instrument for firmness assessment is a penetrometer, which drives a metal plunger into the fruit flesh and records the maximum resistance force. This technique has three main drawbacks [1]: it is destructive, leaving the fruit unsaleable, measurements are highly variable (up to 30%) and it cannot be used in online situations. A fast and nondestructive technique would be desirable for the fresh fruit industry as it offers the benefit of grading and sorting each individual fruit.

Firmness has been a difficult parameter to measure by fruit graders. To date, no commercially successful nondestructive system has been created on a high-speed grader. Most prior research has focused on mechanical methods such as acoustic resonance, impact response, and force-deformation [2–6]. Most of the mechanical methods require contact with the fruit, which limits the grading speed due to the difficulty of achieving reliable physical contact and consistent fruit compliance at high speeds. It also potentially causes physical damage to the fruit. Moreover, mechanical methods are sensitive to each method's specific mechanical property such as deformation force, so they often do not correlate well or consistently with the penetrometer. For this reason, the industry is reluctant to adopt these methods [7]. This has led to more research into the use of optical methods, which have the unique feature of being noncontact. Modern high-speed fruit-grading systems run at speeds in excess of 10 fruit per second and noncontact methods will be advantageous in such circumstances.

This chapter reviews the current optical techniques for firmness measurement. Among these techniques, near-infrared spectroscopy (NIRS) [8–10] and spatially resolved reflectance spectroscopy (SRRS) [11, 12] have been investigated more commonly in recent years, and are more suitable for high-speed operation.

2. Principle of optical methods for measuring firmness

Optical techniques are based on light interactions with fruit tissue. In the visible to near-infrared (Vis/NIR) range of the electromagnetic spectrum, fruit can be considered as semi-transparent or turbid. There are two optical phenomena that describe how light interacts with turbid biological material: absorption and scattering (**Figure 1**). Absorption is primarily due to the chemical composition of the tissue (pigments, chlorophylls, water, etc.). Scattering depends on microscopic changes in refractive index caused by the tissue density, cell composition, and extra- and intra-cellular structure of the fruit, and thus may be useful for assessing textural properties such as firmness. The light transportation in fruit can be characterized by fundamental optical properties of absorption, scattering and refraction, which are defined by the absorption coefficient (μ_a), scattering coefficient (μ_s), refractive index (n), and anisotropy factor (g).

Cen et al. [14] used a hyperspectral backscattering system to measure optical properties of "Golden Delicious" and "Granny Smith" apples over 30 days' storage time. The optical properties from 300 to 1000 nm were compared with acoustic and impact firmness. They found the scattering coefficient generally decreased as the fruit softened ($r > 0.9$ for both mechanical properties). Absorption coefficients also had high correlations with firmness ($r \sim 0.9$ for "Golden Delicious") in the wavelength range that associated with chlorophyll and anthocyanin absorption.

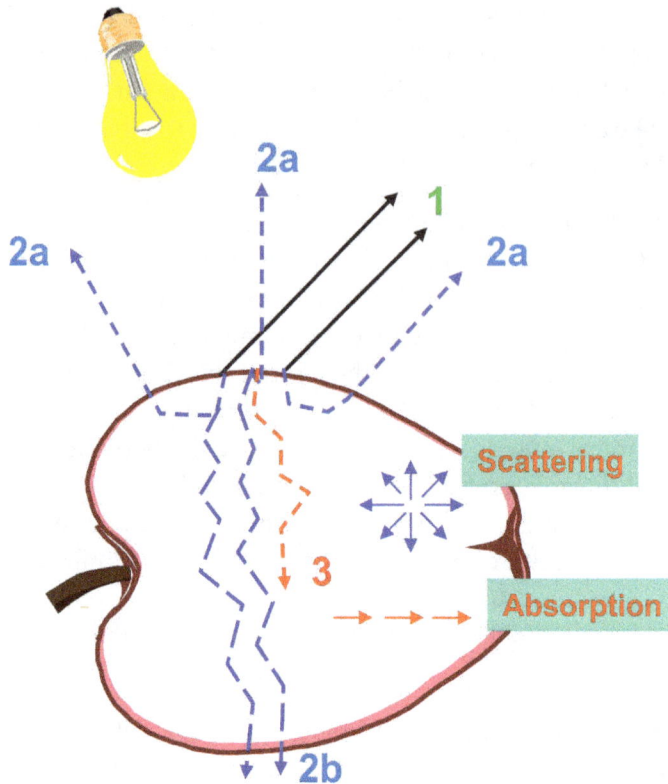

Figure 1. Distribution of incident light in fruits: (1) surface/specular reflectance, (2a) diffuse reflectance, (2b) transmittance, and (3) absorption [13].

The inverse adding-doubling (IAD) technique was used to measure optical properties between 400 and 1050 nm in another study on apples [15]. The reduced scattering coefficient between 550 and 900 nm had an average correlation $r = -0.68$ with penetrometer firmness. Changes in optical properties at carotenoid (400–500 nm) and chlorophyll-a (680 nm) wavelengths correlated with penetrometer firmness with $r = -0.69$ and 0.52, respectively. However, Tomer et al. [16] found that the IAD technique could be quite inaccurate for absorption coefficient measurements on fresh produce at 785 nm, reporting a coefficient for fresh onions that was five times larger than that required for light transport modeling on onions.

There have been some studies based on other optical principles. For example, Costa et al. [17] used a biospeckle laser system to measure the biospeckle images on the *Acrocomia aculeata* fruit pulp. The calculated biological activity (BA) had a negative correlation with penetrometer firmness. The correlation varied depending which tree was evaluated, the highest was $r = -0.903$. Skic et al. [18] used a similar approach for "Ligol" and "Szampion" apples. They achieved a correlation of about $r = -0.5$ for both cultivars. Peña-Gomar et al. [19] used a technique called laser reflectometry near the critical angle (LRCA) to measure the refractive index of mango pulp, which was expected to correlate with acoustic firmness. Their results showed some correlation but the authors did not report the correlation coefficient, and only six fruits were measured.

3. Optical techniques for firmness measurements

Optical methods are noncontact; a feature that distinguishes them from most mechanical methods. In the past two decades, the most common optical sensing method for produce grading is NIRS. Grading lines equipped with NIR sensors are now commercially available from many manufacturers. Firmness is not an attribute commonly assessed using industrial NIR sensors [1], but it has been studied in a number of research applications (**Table 1**).

In theory, the optical scattering properties are more directly related to firmness than absorption properties and have been reported to correlate with firmness, as discussed in Section 2 [14, 15]. Optical techniques that can measure optical properties of biological materials have been studied more recently, aiming to provide a more accurate and robust technique compared to NIRS. These techniques may be divided into three main categories: time resolved, frequency domain, and spatially resolved. Time-resolved and frequency domain techniques have been extensively researched in the biomedical area, but they may not be suitable for applications on a grader line because of expensive instrumentation, slow speed, and the requirement of good contact between the sample and detector [20]. Spatially resolved techniques, and more specifically SRRS, have been researched more commonly for such applications as it can overcome many of those deficiencies.

3.1. Near-infrared spectroscopy

NIRS is widely used to determine fruit quality parameters, particularly compositional parameters such as soluble solids or dry matter content [4, 21]. Standard NIRS measures the spectral pattern of light transmitted through a representative portion of the flesh, and chemometric analysis methods are generally used to interpret the resulting absorbance spectra in terms of the parameters of interest. The disadvantage is that this technique relies on a prior extensive training exercise to develop a predictive model, based on the careful selection and measurement of a representative calibration data set from a suitable population. The model also needs to be checked and updated constantly.

Species	Cultivar	Acquisition mode	Spectral range	Prediction	Reference
Apple	"Royal Gala"	Interactance	500–1100 nm	$R = 0.77$ RMSEP = 7.5 N	McGlone et al. [25]
Apple	"Gala" "Red Delicious"	Reflectance	400– 1800 nm	$R = 0.88$	Park et al. [26]
Mandarin	Satsuma	Reflectance	350–2500 nm	$R = 0.83$	Gómez et al. [10]
Pear	"Conference"	Reflectance	780– 1700 nm	$R = 0.59$	Nicolaï et al. [9]
Capsicum annuum	Bell pepper	Reflectance	780– 1690 nm	$R = 0.6$ RMSEP = 7.2 N	Penchaiya et al. [8]
Cherry	"Hedelfinger Sam"	Reflectance	800–1700 nm	$R = 0.8$ & 0.65 SEP = 0.79 and 0.44 N	Lu [27]

Table 1. Overview of applications of NIR spectroscopy in firmness measurements.

For measuring fruit firmness, the NIRS method is limited in theory because it involves measurement of the apparent light absorbing power of a sample, which does not segregate scattering and absorption properties. However previous studies have suggested firmness may affect the apparent light-absorbing power through chemical changes associated with cell wall degradation, physical changes in intercellular structure and/or indirectly through correlated pigment absorption changes such as a chlorophyll decrease on ripening [14, 15].

3.1.1. Basic concepts

Near-infrared radiation covers the range of the electromagnetic spectrum between 780 and 2500 nm. Often wavelengths below 780 nm are also included in the analysis as these regions contain valuable information on absorbing pigments within the fruit flesh and skin [15]. Therefore, this technique is often referred to as Vis/NIR spectroscopy.

The typical NIRS set-up uses a broadband light source to illuminate the sample and the transmitted or reflected light is measured using a spectrometer. In the design process, it is useful to know that the NIR light intensity decreases exponentially with depth. One study [22] showed that the light intensity dropped to 1% of the initial intensity at a depth of 25 mm inside an apple in the 700–900 nm range. The depth was less than 1 mm in the 1400–1600 nm range. Therefore, the optical arrangement and the effective optical path length for the light are crucial elements to consider in order to collect spectra containing relevant information from the sample. This also explains why NIRS is suited for use with thin-skinned fruit, the thicker skins limiting light penetration [23].

In practice, three measurement set-ups are used (**Figure 2**). In reflection mode, light source and spectrometer are on one side but at a specific angle to avoid specular reflection, while in

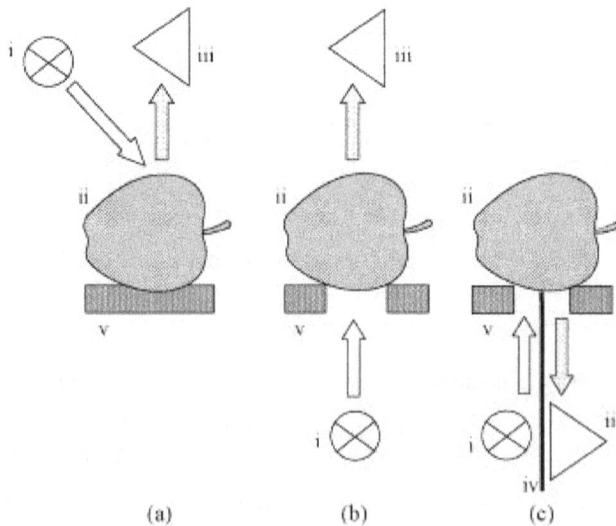

Figure 2. Three different set-ups: (a) reflectance, (b) transmittance, and (c) interactance. (i) Is the light source, (ii) is the sample, (iii) is the detector, (iv) is a light barrier, and (v) is the mechanical support [21] (Used with permission from Elsevier).

transmission mode the light source and detector are on opposite sides. Interactance requires a special optical arrangement so that specular and surface reflection cannot directly enter the detector.

Transmission measurement has the advantages of exploring the largest volume of the internal flesh and all the light measured has interacted with the flesh. Thus it is suitable to find internal defects, but the transmitted light might also contain information of the two layers of skin (front entrance and back exit), and the core of the fruit. For firmness measurement, although light penetration is limited and one skin layer is still present, reflection and interactance set-ups will be more desirable as the light interacts with some portion of flesh without interference from the core. Schaare and Fraser [24] compared reflectance, interactance and transmission measurements for measuring soluble solid content (SSC), density and internal flesh color of kiwifruit and concluded that interactance measurements provided the most accurate results.

3.1.2. Firmness applications

Sensors based on NIRS techniques have been mainly developed for chemical compositions such as SSC, and most of the studies have been carried out under static conditions. The industry is taking the lead in the development of online systems, but there is little scientific evidence of their accuracies [21]. Attempts to use NIRS for fruit firmness prediction have met with varying degrees of success with some studies reporting correlations as high as $R \sim 0.8$ – 0.9. **Table 1** gives an overview of NIRS applications that measure firmness of fruits and vegetables.

Most reported scientific studies consider only a single NIR instrument format for fruit assessment. For example, McGlone et al. [25] used an interactance mode (**Figure 3**). The system contained a broadband light source (50 W quartz halogen, RJL 5012 FL, Radium, Germany) and a nonscanning polychromatic diode array spectrometer (Zeiss MMS1-NIR, Germany). Fruits were placed on a holder with stem-calyx horizontal. Measurements were generally taken on two opposite sides around the circumference, taking care to avoid any obvious surface defects. The absorbance spectrum measured was the average of 5 contiguous acquisitions at 175 ms integration time.

The wavelength range used varies among the reported literature studies (**Table 1**). Walsh [23] suggested that restricted wavelength ranges could improve the robustness of a model and allow for the development of lower cost "multispectral" measurement systems. Prediction performance was generally determined by dividing the fruits randomly into a calibration and a validation set for model development. Walsh [23] also reported that such a model will predict the attribute of interest within the population, but it is likely to fail spectacularly on a new, independent set.

3.2. Spatially resolved reflectance spectroscopy (SRRS)

Figure 1 illustrates two types of light reflectance: surface reflectance and diffuse reflectance. Surface reflectance contains information about the object surface such as color. Only 4–5%

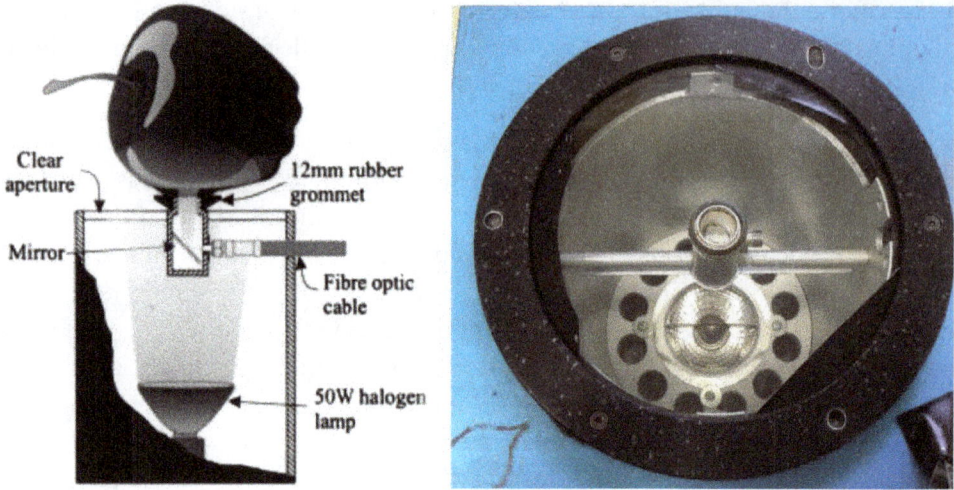

Figure 3. The benchtop NIRS system [25].

of incident light is reflected by surface reflection and external diffuse reflectance, so most reflected light contains the diffuse reflected/backscattered photons that carry information of the internal tissue properties [11].

Figure 4 shows a small continuous-wave light beam perpendicularly illuminates the sample's surface, and the reflected light is measured at different distances from the light source, forming the spatial profile (**Figure 3**). Optical properties/parameters can be obtained by using a phenomenological diffusion model and/or a heuristic modified Lorentzian model from the

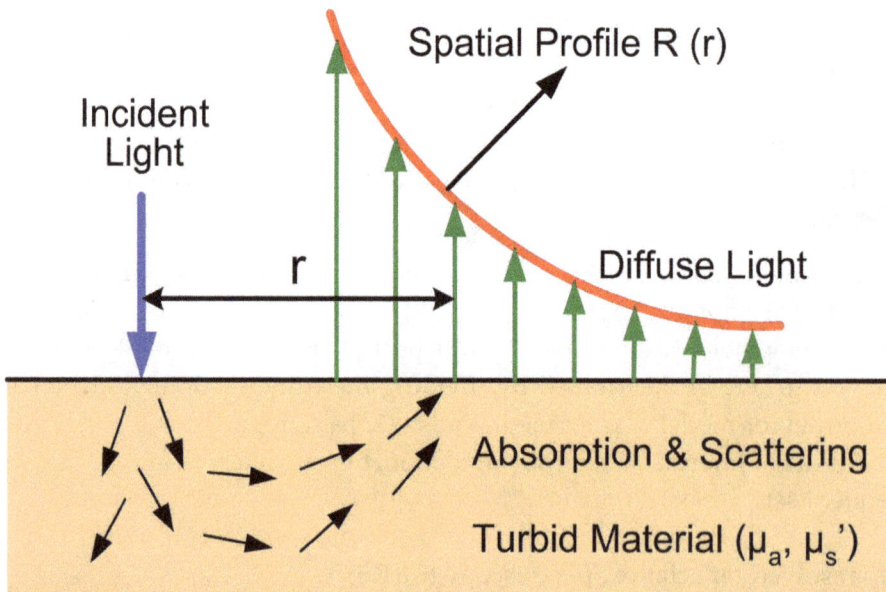

Figure 4. Measuring principle for spatially resolved reflectance spectroscopy (SRRS) [20] (Used with permission from the author).

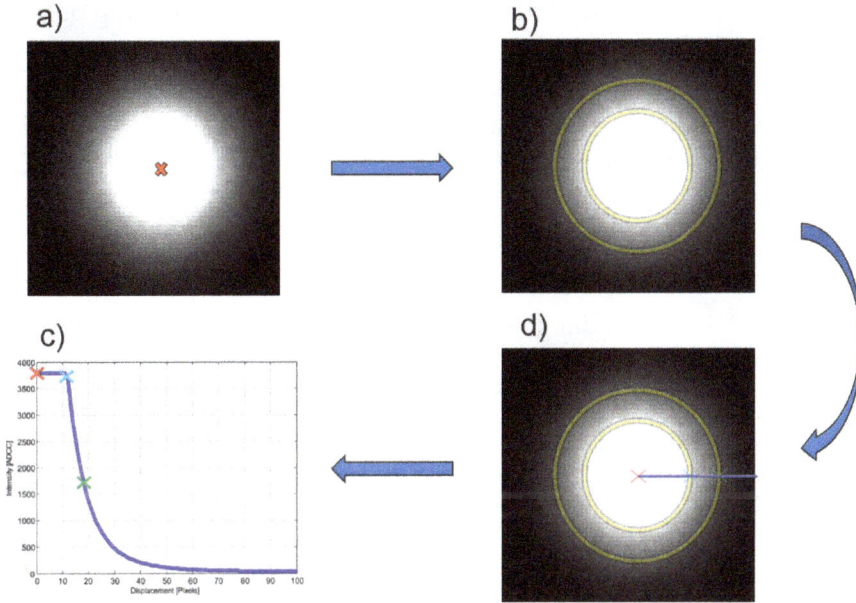

Figure 5. Imaging processing used by Sun et al. [30] for apple firmness measurements: (a) finding center in the raw image, (b) ring average, and (c) & (d) producing the spatial profile.

measured one-dimensional scattering profile. Mollazade et al. [28] used texture-based features methods to build models to predict mechanical properties of various produce. Instead of looking at a single 1D scattering profile, this technique analyzed the entire 2D images, which was expected to improve the correlation to firmness.

The extracted parameters can then be used to predict firmness using statistical models such as multiple linear regression (MLR) and artificial neural network (ANN). Typically, images are first processed to reduce noise and then converted into one-dimensional profile [29]. **Figure 5** illustrates the process used by Sun et al. [30] for measuring apple firmness. The scattering image was first processed to find the center of the illuminated area (**Figure 5(a)**). Then a process called ring/radial averaging was performed. The distance to each pixel was calculated and rounded to the nearest whole number (**Figure 5(b)**). All pixels at each of these integer radii were grouped and averaged providing a vector of intensity values that correspond to single pixel rings expanding out from the center point (**Figure 5(d)**). The intensity profile (**Figure 5(c)**) was finally produced.

3.2.1. Parameters extraction

In turbid material, a diffusion equation is often used as an approximation of the transport of the light. For SRRS under the assumption of inexistence of photon source in the medium, the diffusion equation can be simplified to an equation consisting of three variables: r (source-detector distance), μ_a and $\mu_{s'}$ [11, 31]. Unknown optical properties μ_a and $\mu_{s'}$ can be obtained by applying a curve fitting procedure with respect to r.

Researchers have also used statistical distribution functions to fit scattering profiles as a function of scattering distance. Peng and Lu [32] investigated a number of variations of modified Lorentzian functions aiming to find one suitable for firmness and SSC measurements. They concluded Eq.(1) was the best performing equation, which was also used in other studies for firmness applications [7, 29, 30]:

$$I(x) = a + \frac{b}{1 + \left(\frac{|x|}{c}\right)^d} \tag{1}$$

where I is the intensity along a radial intensity profile, a is the asymptotic value of light intensity when x (distance to center of the light spot) approaches infinity, b is the peak value corresponding to the intensity at the center of the image, c is the full width half maximum (FWHM) of the intensity profile, and d is related to the slope of the profile in the FWHM region.

3.2.2. Hardware

A SRRS system consists of two essential components: light source and imaging system. All the systems can be divided into three types according to the light source and operating wavelength range: laser light backscatter imaging (LLBI), multispectral light backscatter imaging (MLBI), and hyperspectral light backscatter imaging (HLBI).

The LLBI technique requires a small illumination spot on the target fruit, and measurement scattering areas of 25–30 mm diameter have been used for beam diameters of 0.8–1.5 mm by Lu [33] and Peng and Lu [32], respectively. Lasers are particularly suitable for this purpose since lasers can produce focused high-irradiance illumination spots on the fruit, which allows for deeper light penetration and fast image acquisition (shorter integration time). Moreover, LLBI systems are more robust and cost-effective than MLBI and HLBI. Overall, LLBI systems are potentially suitable for online high-speed operations. One of the drawbacks of LLBI systems is the limited operating wavelength. One to four lasers are typically used [28, 30, 34].

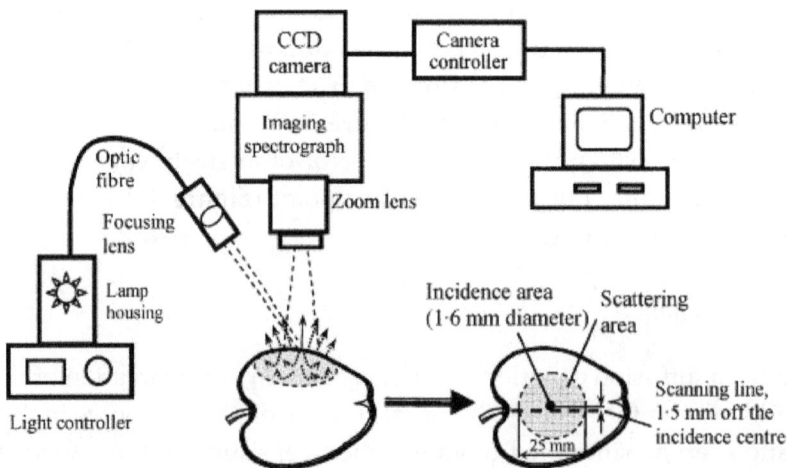

Figure 6. Hyperspectral system (HLBI) for measuring the firmness of peach [36] (Used with permission from Elsevier).

Species	Cultivar	Light source	Detector	Spectral range	Prediction	Reference
Apple	"Golden Delicious" "Jonagold Delicious"	Halogen lamp	CCD camera with spectrograph	500–1000 nm	$r = 0.84 – 0.95$ SEP = 5.9 – 8.7 N	Mendoza et al. [37]
Apple	"Golden Delicious"	Halogen lamp	CCD camera with spectrograph	450–1000 nm	$r = 0.894$ SEP = 6.14 N	Peng and Lu [32]
Apple	"Braeburn"	Halogen lamp	CCD camera with spectrograph	500–1000 nm	$r = 0.84$ RMSEP = 9.68 N	Nguyen Do Trong et al. [38]
Apple	"Braeburn"	Supercontinuum laser with monochromator	CCD camera	550–1000 nm	$r = 0.8$ RMSEP = 0.75 kg cm^{-2}	Van Beers et al. [35]
Peach	"Red Haven" "Coral Star"	Halogen lamp	CCD camera with spectrograph	500–1000 nm	$r = 0.76 – 0.88$	Lu and Peng [36]
Apple	"Red Delicious" "Golden Delicious"	Quartz tungsten halogen lamp	NIR enhanced CCD camera with a liquid crystal tunable filter	680, 700, 740, 800, 820, 910, & 990 nm	$r = 0.898$ SEV = 6.14 – 6.41 N	Peng & Lu [29]
Apple	"Elstar" "Pinova"	Laser diode	CCD camera	680,780,880,940, & 980 nm	$r = 0.89 – 0.81$ RMSEC $V = 4.71 –$ 5.48 N/cm^2	Qing et al. [34]
Apple	"Royal Gala"	Laser diode	CMOS camera	685,850,904, & 980 nm	$r = 0.87$ RMSECV = 7.17 N	Sun et al. [30]
Apple Plum Tomato mushroom	"Pinova" & "Elstar" "Jojo" & "Tophit" "Pannovy"	Laser diode	CCD camera	660 nm	$r = 0.887 – 0.919$	Mollazade et al. [28]
Apple	"Golden Delicious"	Laser diode	CCD camera with spectrograph	408 nm	$R = 0.75$ SEP = 8.57 N	Noh & Lu [39]

Table 2. Overview of applications of SRRS in firmness measurements.

In the MLBI and HLBI systems, the light source is a tungsten-halogen lamp. The light usually passes through an optical fiber and then focuses on the fruit by a collimating lens, as shown in **Figure 6**. One exception is the system developed by Van Beers et al. [35] where a super-continuum laser and a monochromator were used for the hyperspectral measurements.

The scattering profiles can be measured using multiple spectrometers at different source-detector distances. The advantage of using a spectrometer is that multiple wavelengths or a specific spectral region can be obtained simultaneously. However, it requires a good contact/focus between the probes and the sample, which will not be suitable for online operations. A CCD camera is more commonly used as it is noncontact, which has been a dominant format in all three types of systems (**Table 2**), except that Sun et al. [30] used a CMOS camera. CCD and CMOS cameras allow only single wavelength operation, but an imaging spectrograph has been used in HLBI systems to provide spectral and spatial information on a single image (**Figure 6**). Filters were also used in MLBI system to enable the image acquisition at specific wavelength [33].

3.2.3. Applications

An overview of SRRS to measure the firmness of fruits and vegetables is given in **Table 2**. The studies show that SRRS achieves similar performance compared with NIRS. The correlations with penetrometer firmness are often in the range of $r = 0.8 - 0.9$. It is not clear which type or instrument format of SRRS is more advantageous. Most studies evaluated the potential of SRRS systems for firmness measurements on static fruit and have not considered the practical challenges of applying SRRS to online situations. Unlike NIRS, there have been no commercially available sensors based on SRRS. All the studies listed in **Table 2** are laboratory systems specifically constructed for measuring stationary fruits. Lu and Peng [7] developed a real-time LLBI system for measuring the firmness of apples on a belt conveyor and achieved a correlation of $r = 0.86$. They claimed that the LLBI system could be integrated into existing grader lines without significant modification. However, their measurements were taken when the conveyor speed was only two fruit per second which is well below the maximum speed of a modern grader. Also, the fruit was manually positioned so that the scattering images could be captured from the equatorial areas of the fruit. The authors suggested the lasers and CCD camera should allow faster acquisition of the scattering images, but the algorithm for processing the images was the bottleneck. Overall, fruit orientation and data processing speed are the main challenges for applying SRRS in online systems.

4. Conclusion

For the main two optical techniques discussed here, NIRS and SRRS, there have been prior studies showing correlations with penetrometer firmness as high as $r = 0.8 - 0.9$. Both techniques can come in many instrument formats, so it is hard to judge from the literature which instrument is more advantageous. A direct comparison of the NIRS and SRRS methods has not been performed on the exact same fruit samples commonly. Sun et al. [30] compared an interactance mode NIRS system with an LLBI system using "Royal Gala" apples. The two systems had similar correlations with penetrometer firmness of about $r = 0.9$. By contrast, a comparison of a reflectance mode NIRS system and an MLBI system using "Red Delicious" and "Golden Delicious" was conducted by Lu and Peng [40]. Their MLBI system outperformed NIRS system with $r = 0.82$ and 0.81 for two apple cultivars, versus $r = 0.5$ and 0.48 from the NIRS system.

It has been suggested $r = 0.94$ ($r^2 = 0.89$) be considered as a minimum for any useful sorting/grading purposes [41]. Although sometimes very close to that mark, the correlations reported here and in most previous studies are lower. Moreover, NIRS sensors are likely to perform worse across grader lines and seasons because of the low robustness of the calibration models. These may explain why there are no optical sensors for firmness measurements yet commercially available. For SRRS, another concern is the feasibility of online applications; most studies discussed here are bespoke laboratory systems for measuring static fruits. Fruit speed and orientation are normally not a problem for NIRS but might be an issue for the online application of SRRS.

NIRS is a relatively mature technique for quality grading of fruits and vegetables, though not commonly used for firmness. SRRS might well be a better method for firmness, being more robust in practice as it is more directly linked to the optical scattering properties that are presumed to be directly affected by changes in texture properties. However, the SRRS systems will have to be improved and demonstrate better performance than has been achieved to date before they can be considered for commercial implementation. We recommend further research across a wider variety of fruits in the future, and feasibility studies to assess the potential of SRRS for online applications.

Author details

Jason Sun[1,2]*, Rainer Künnemeyer[1,2] and Andrew McGlone[3]

*Address all correspondence to: zhesun89@gmail.com

1 School of Engineering, University of Waikato, Hamilton, New Zealand

2 Dodd Walls Centre for Photonic and Quantum Technologies, New Zealand

3 The New Zealand Institute for Plant & Food Research, Hamilton, New Zealand

References

[1] García-Ramos FJ, Valero C, Homer I, Ortiz-Cañavate J, Ruiz-Altisent M. Non-destructive fruit firmness sensors: A review. Spanish Journal of Agricultural Research. 2005;3(1):61-73

[2] Abbott JA. Quality measurement of fruits and vegetables. Postharvest Biology and Technology. 1999;15(3):207-225

[3] De Ketelaere B, Howarth MS, Crezee L, Lammertyn J, Viaene K, Bulens I, De Baerdemaeker J. Postharvest firmness changes as measured by acoustic and low-mass impact devices: A comparison of techniques. Postharvest Biology and Technology. 2006;**41**(3):275-284

[4] Ruiz-Altisent M, Ruiz-Garcia L, Moreda GP, Lu R, Hernandez-Sanchez N, Correa EC, Diezma B, Nicolaï B, García-Ramos J. Sensors for product characterization and quality of specialty crops—A review. Computers and Electronics in Agriculture. 2010;**74**(2):176-194

[5] Steinmetz V, Crochon M, Bellon Maurel V, Garcia Fernandez JL, Barreiro Elorza P, Verstreken L. Sensors for fruit firmness assessment: Comparison and fusion. Journal of Agricultural Engineering Research. 1996;**64**(1):15-27

[6] Hitchman S, van Wijk K, Davidson Z. Monitoring attenuation and the elastic properties of an apple with laser ultrasound. Postharvest Biology and Technology. 2016;**121**:71-77

[7] Lu R, Peng Y. Development of a multispectral imaging prototype for real-time detection of apple fruit firmness. Optical Engineering. 2007;**46**(12):123201

[8] Penchaiya P, Bobelyn E, Verlinden BE, Nicolaï BM, Saeys W. Non-destructive measurement of firmness and soluble solids content in bell pepper using NIR spectroscopy. Journal of Food Engineering. 2009;**94**(3-4):267-273

[9] Nicolaï BM, Verlinden BE, Desmet M, Saevels S, Saeys W, Theron K, Cubeddu R, Pifferi A, Torricelli A. Time-resolved and continuous wave NIR reflectance spectroscopy to predict soluble solids content and firmness of pear. Postharvest Biology and Technology. 2008; **47**(1):68-74

[10] Gómez AH, He Y, Pereira AG. Non-destructive measurement of acidity, soluble solids and firmness of satsuma mandarin using Vis/NIR-spectroscopy techniques. Journal of Food Engineering. 2006;**77**(2):313-319

[11] Mollazade K, Omid M, Tab FA, Mohtasebi SS. Principles and applications of light backscattering imaging in quality evaluation of agro-food products: A review. Food and Bioprocess Technology. 2012;**5**(5):1465-1485

[12] Adebayo SE, Hashim N, Abdan K, Hanafi M. Application and potential of backscattering imaging techniques in agricultural and food processing – A review. Journal of Food Engineering. 2016;**169**:155-164

[13] Xie L, Wang A, Xu H, Fu X, Ying Y. Applications of near-infrared systems for quality evaluation of fruits: A review. ASABE. 2016;**59**(2):399-419

[14] Cen H, Lu R, Mendoza F, Beaudry RM. Relationship of the optical absorption and scattering properties with mechanical and structural properties of apple tissue. Postharvest Biology and Technology. 2013;**85**:30-38

[15] Rowe PI, Künnemeyer R, McGlone A, Talele S, Martinsen P, Seelye R. Relationship between tissue firmness and optical properties of Royal Gala apples from 400 to 1050 nm. Postharvest Biology and Technology. 2014;**94**:89-96

[16] Tomer N, McGlone A, Künnemeyer R. Validated simulations of diffuse optical transmission measurements on produce. Computers and Electronics in Agriculture. 2017;**134**:94-101

[17] Costa AG, Pinto FA, Braga RA, Motoike SY, Gracia L. Relationship between biospeckle laser technique and firmness of *Acrocomia aculeata* fruits. Revista Brasileira de Engenharia Agrícola e Ambiental. 2017;**21**(1):68-73

[18] Skic A, Szymańska-Chargot M, Kruk B, Chylińska M, Pieczywek PM, Kurenda A, Zdunek A, Rutkowski KP. Determination of the optimum harvest window for apples using the non-destructive biospeckle method. Sensors. 2016;**16**(5):661

[19] Peña-Gomar M, Arroyo-Correa G, Aranda J, editors. Characterization of Firmness Index for Mango Fruit by Laser Reflectometry around the Critical Angle. XIth International Congress and Exposition; USA: Society for Experimental Mechanics Inc;2008

[20] Cen H. Hyperspectral Imaging-Based Spatially-Resolved Technique for Accurate Measurement of the Optical Properties of Horticultral Products. Michigan, USA: Michigan State University; 2011

[21] Nicolaï BM, Beullens K, Bobelyn E, Peirs A, Saeys W, Theron KI, Lammertyn J. Nondestructive measurement of fruit and vegetable quality by means of NIR spectroscopy: A review. Postharvest Biology and Technology. 2007;**46**(2):99-118

[22] Fraser DG, Künnemeyer R, McGlone VA, Jordan RB. Near infra-red (NIR) light penetration into an apple. Postharvest Biology and Technology. 2001;**22**(3):191-195

[23] Walsh K. Nondestrutive assessment of fruit quality. In: Wills RB, Golding J, editors. Advances in Postharvest Fruit and Vegetable Technology. USA: CRC Press; 2015. pp. 39-64

[24] Schaare PN, Fraser DG. Comparison of reflectance, interactance and transmission modes of visible-near infrared spectroscopy for measuring internal properties of kiwifruit (*Actinidia chinensis*). Postharvest Biology and Technology. 2000;**20**(2):175-184

[25] McGlone VA, Jordan RB, Martinsen PJ. Vis/NIR estimation at harvest of pre- and post-storage quality indices for Royal Gala apple. Postharvest Biology and Technology. 2002;**25**(2):135-144

[26] Park B, Abbott JA, Lee KJ, Choi CH, Choi KH. Near-infrared diffuse reflectance for quantitative and qualitative measurement of soluble solids and firmness of delicious and gala apples. Transactions of the ASAE. 2003;**46**(6):1721-1731

[27] Lu R. Predicting firmness and sugar content of sweet cherries using near-infrared diffuse reflectance spectroscopy. Transactions-American Society of Agricultural Engineers. 2001;**44**(5):1265-1274

[28] Mollazade K, Omid M, Akhlaghian Tab F, Kalaj YR, Mohtasebi SS, Zude M. Analysis of texture-based features for predicting mechanical properties of horticultural products by laser light backscattering imaging. Computers and Electronics in Agriculture. 2013;**98**:34-45

[29] Peng Y, Lu R. Improving apple fruit firmness predictions by effective correction of mul-
      tispectral scattering images. Postharvest Biology and Technology. 2006;**41**(3):266-274

[30] Sun J, Künnemeyer R, McGlone A, Rowe P. Multispectral scattering imaging and NIR
      interactance for apple firmness predictions. Postharvest Biology and Technology. 2016;
      119:58-68

[31] Cen H, Lu R, Dolan K. Optimization of inverse algorithm for estimating the optical
      properties of biological materials using spatially-resolved diffuse reflectance. Inverse
      Problems in Science and Engineering. 2010;**18**(6):853-872

[32] Peng Y, Lu R. Analysis of spatially resolved hyperspectral scattering images for assess-
      ing apple fruit firmness and soluble solids content. Postharvest Biology and Technology.
      2008;**48**(1):52-62

[33] Lu R. Multispectral imaging for predicting firmness and soluble solids content of apple
      fruit. Postharvest Biology and Technology. 2003;**31**(2):147-157

[34] Qing Z, Ji B, Zude M. Non-destructive analyses of apple quality parameters by means
      of laser-induced light backscattering imaging. Postharvest Biology and Technology.
      2008;**48**(2):215-222

[35] Van Beers R, Aernouts B, Gutiérrez LL, Erkinbaev C, Rutten K, Schenk A, Nicolaï B,
      Saeys W. Optimal illumination-detection distance and detector size for predicting brae-
      burn apple maturity from Vis/NIR laser reflectance measurements. Food and Bioprocess
      Technology. 2015; **8**(10):2123-2136

[36] Lu R, Peng Y. Hyperspectral scattering for assessing peach fruit firmness. Biosystems
      Engineering. 2006;**93**(2):161-171

[37] Mendoza F, Lu R, Ariana D, Cen H, Bailey B. Integrated spectral and image analysis of
      hyperspectral scattering data for prediction of apple fruit firmness and soluble solids
      content. Postharvest Biology and Technology. 2011;**62**(2):149-160

[38] Nguyen Do Trong N, Erkinbaev C, Tsuta M, De Baerdemaeker J, Nicolaï B, Saeys W.
      Spatially resolved diffuse reflectance in the visible and near-infrared wavelength range
      for non-destructive quality assessment of braeburn apples. Postharvest Biology and
      Technology. 2014;**91**:39-48

[39] Noh HK, Lu R. Hyperspectral laser-induced fluorescence imaging for assessing apple
      fruit quality. Postharvest Biology and Technology. 2007;**43**(2):193-201

[40] Lu R, Peng Y. Comparison of multispectral scattering and Visible/NIR spectroscopy for
      predicting apple fruit firmness. Information and Technology for Sustainable Fruit and
      Vegetable Production, (FRUTIC). 2005;**5**:493-502

[41] McGlone VA, Kawano S. Firmness, dry-matter and soluble-solids assessment of postharvest
      kiwifruit by NIR spectroscopy. Postharvest Biology and Technology. 1998;**13**(2):131-141

3

Postharvest Handling of Berries

Sandra Horvitz

Additional information is available at the end of the chapter

Abstract

Strawberries, raspberries, and blackberries are highly appreciated fruits due to their unique taste and high content in antioxidant and bioactive compounds. They are rich in phenolic compounds, mostly flavonoids and anthocyanins, which are responsible for fruit color and can exert antioxidant, antimicrobial, anti-inflammatory, anticancer, and cardioprotective effects. However, berries have a short storage life, as a result of their high respiration and softening rate, and susceptibility to mechanical damages and decay. As berries are considered non-climacteric fruit, they must be harvested at, or near to full maturity, because they will not continue to ripen normally once detached. At this stage, the fruit presents appropriate organoleptic attributes but may become softer and more sensitive to mechanical damage. Thus, it is crucial to be extremely careful during harvest and postharvest handling and to sort, grade, and pack the berries in the field, avoiding excessive manipulation of the fruit. The most extended methods to maintain quality during the postharvest period are prompt precooling and storage at low temperatures. Modified and controlled atmospheres with up to 20-kPa CO_2 and 5–10-kPa O_2 reduce microbial growth and delay senescence but can affect bioactive compounds with a cultivar-dependent response observed for these technologies.

Keywords: berries, maturity index, packaging, refrigeration, storage

1. Introduction

Berry fruits include, among others, strawberries (*Fragaria ananassa*), raspberries (*Rubus idaeus*), and blackberries (*Rubus* spp.). These fruits are characterized by their acidic taste and can be consumed fresh or frozen. Fresh fruits are mainly consumed locally and are available only in the ripening season, except countries from South America, like Colombia or Ecuador, where the production occurs all year round. Berries are also available as processed products like refrigerated fruit pulp, jams, juices, and nectars [1]. What's more, due to their high content in antioxidant and bioactive compounds, they can be considered as functional foods. In effect, different studies conducted on berry fruits report antioxidant, antimicrobial, anti-inflammatory, anticancer, and cardioprotective effects, which were attributed to their high content in bioactive compounds, mainly different phenolic compounds [2].

However, the manipulation of these fruits presents a series of challenges: berries lack a protective peel and are highly perishable, mainly because of their susceptibility to mechanical damage, water loss, and fungal decay [3]. What's more, berries are considered non-climacteric fruit, which implies that they need to be harvested at, or near to, full maturity as most of them will not continue to ripen normally once detached, and eating quality will not improve after harvest. In some cases, they can color in storage but if they are harvested too early, texture, sweetness, and acidity fail to fully develop [4].

Fruit quality for the market is largely determined by physicochemical parameters like size, full color, gloss, firm and crisp texture, absence of decay, injuries and bruises, a balance between sweetness and acidity, green sepals, and typical aroma. At the same time, the main causes of loss and rejects include weight loss, presence of bruises and cuts, symptoms of mold and decay, color changes, juice leakage, and sepal wilt.

To get the maximum quality at harvest and maintain this quality during transport and commercialization until the fruit is consumed, it is essential to harvest berries at the optimum stage of maturity [5]. In this sense, the UNECE Standard FFV-57 [6] concerning the marketing and commercial quality control of berry fruit establishes that "Berry fruits must be sufficiently developed and display satisfactory ripeness according to the species but must not be overripe," emphasizing the need to harvest at the appropriate maturity stage for each type of fruit.

2. Harvest

In order to avoid excessive manipulation and damage to the fruit, berries for the fresh market should be hand-harvested, sorted, graded, and packed in the field, directly into the final container. Fruit ripeness at harvest and fruit handling are two critical factors in the postharvest keeping quality. In fact, the stage of maturity at harvest largely affects the shelf-life of berries, their storage behavior, and sales probability [7]. Immature fruit may have a longer storage capability but are unlikely to develop appropriate organoleptic characteristics while shelf-life of over-mature fruit is generally very short as the susceptibility to decay also increases [8].

As berries ripen quickly but non-uniformly (**Figure 1**), it is crucial to harvest frequently (daily, or every 2–3 days, depending on weather conditions and area of production) and also train pickers to identify the proper ripening stage and in the correct harvest practices to avoid damages to the fruit.

Ideally, the fruit should be harvested early in the morning, after the dew is off the berries or in the evening when the temperatures are cooler [9]. Berries should not be touched before harvest because they are extremely fragile and easily damaged during harvest, for example, by finger pressure. Only sound berries with good appearance should be placed in the packages,

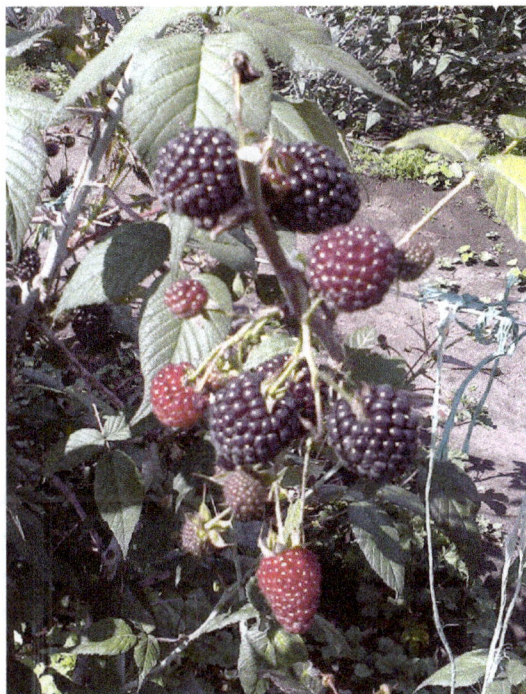

Figure 1. Fruit of blackberries showing different maturity stages.

and once harvested, fruit must be protected from exposition to direct sunlight. Rotten fruit must be picked off plants and discarded far from the marketable berries to avoid contamination of the latter while small and overripe fruits can be used for processing [10].

2.1. Harvest maturity

Different maturity indexes can be used for determining berries' optimum harvest date. However, harvest maturity is mainly determined by fruit surface color and most standards require for strawberries that more than one-half to three-fourth of the surface to be colored. In the case of raspberries and blackberries, the fruit must present a completely red and a bright, dark purple/black color, respectively. Color is also the main criterion used by the consumer to judge fruit quality [11]. Besides color, blackberries, and raspberries should pull easily from the receptacle yet being still firm. Regardless of the berry considered and in addition to color, appearance (size, shape, and absence of defects), firmness, flavor (soluble solids, titratable acidity, and flavor volatiles), and nutritional value (vitamin C) are all important quality characteristics that must be considered.

Several studies have shown that the color of these berries can change during storage even if the fruit are harvested at early stages of color development [12–14]. However, the changes in sugar and acid content of these unripe fruits are not enough to make them suitable for fresh consumption [4]. On the other hand, Krüger et al. [7] reported that suitability for selling raspberries declined rapidly with increased ripening stage, and thus, the fruit should not be stored and must be sold and consumed immediately after picking.

In any case, it is recommended to avoid mixing different ripening stages in the same pack, as this practice is usually rejected by consumers at the marketplace. At an industrial level, fruit selection is based both on external attributes such as intense red color and color distribution, fruit size and shape, and absence of physiological defects and on internal quality parameters including sweetness, acidity, and flavor [15]. For an acceptable flavor, a minimum of 7% soluble solids and/or a maximum of 0.8% titratable acidity are highly recommended for strawberries [16]. Similarly, the Ecuadorian Quality Standard NTE-2427 [17] for Andean blackberries (*Rubus glaucus* Benth) establishes a minimum of 9% soluble solids, a maximum of 1.8% titratable acidity and a minimum of 5 for the maturity index (total soluble solids/titratable acidity).

2.2. Packing

The containers most commonly used at the supermarket for raspberries and blackberries are plastic clamshells containing 250 g of fruit. In the case of strawberries also, containers for 500, 1000, and even 2000 g of fruit are used, depending on fruit size (**Figure 2a** and **b**). Pulp and wooden containers are also used, but they present the disadvantage that stain easily, and wooden containers are also expensive. Regardless of the material, wide and shallow containers are preferred to deep containers and no more than three layers of fruit should be included in each package as the fruit in the bottom may be crushed by the fruit on top.

Figure 2. Plastic clamshells of 250 (a) and 500 g (b) for the packing of raspberries and blackberries, and strawberries, respectively.

Plastic clamshells present some advantages: they are solid and thus give protection to the fruit from mechanical damage, they do not stain, they are inexpensive, and, as they are usually clear or transparent, they allow consumers to inspect all the fruit at the time of purchasing. The containers must be vented on top and sides and have a lid to reduce mechanical damage and moisture loss. On the other hand, the main disadvantage of these packages is plastic disposal after use.

In some countries, it is still also a very common practice the use of baskets or buckets, sometimes containing up to 12–15 kg of fruit (**Figure 3**). Customers pay for the first basket and bring them back in following purchases. This kind of containers are not appropriate as they are usually not washed or disinfected before reuse and thus lack hygiene and can accumulate fungal spores. Furthermore, the excessive weight of fruit causes damage to the fruit located at the bottom which usually collapses. In order to absorb the juice leaked from the fruit, non-food grade, periodic paper is frequently added at the bottom of the baskets, underneath the fruit, worsening the situation. Some efforts are being made to eliminate these kinds of practices and to replace these containers by cartons containing plastic clamshells or cardboard boxes (**Figure 4a** and **b**).

Figure 3. Baskets used for blackberries harvest.

Figure 4. Cartons containing plastic clamshells of 250 (a) and 500 g (b), for raspberries/blackberries and strawberries, respectively.

3. Precooling

Precooling, consisting in rapid removal of field heat immediately after harvest, is essential to maintain fruit quality and control decay [18]. For example, strawberries rapidly cooled down to 0°C showed threefold the storage life of those fruit maintained at 10°C [19]. Field heat is often removed using forced air cooling, where rapidly moving cold air is forced through pallets of fruit to lower fruit temperature to 0 to 1°C within 2 h of picking. This method is preferred to room cooling, as forced air can cool berries to 1°C within an hour, whereas room cooling may take up to 9 h [9].

High relative humidity (85–95%) should be maintained within the refrigerated rooms, but free moisture on the berries or in the containers must be kept to a minimum as, to reduce fruit rot, the berries must be kept dry. Precooling conditions for blackberries are forced air cooling to 5°C within 4 h and fruit should be transported at refrigeration temperatures of 5°C or less. Raspberries should be forced air cooled to 1°C, no later than 12 h after harvest.

4. Storage

Strawberries, raspberries, and blackberries are highly perishable due to their relatively high water content, high physiological postharvest activity, and susceptibility to fruit rot and darkening. The high respiration rates of these fruits cause changes in texture, color, flavor, and nutritional content during storage, and such changes are crucial for the determination of fruit quality and consumer's acceptability [20]. Their short storage life is also the result of decay caused by rot-causing pathogens and quick softening rates [21]. *Botrytis cinerea*, a necrotrophic fungus that causes gray mold rot, is one of the main pathogens responsible for postharvest decay in berries. The symptoms of disease are soft rot with a collapse and water soaking of parenchyma tissues, followed by the appearance of gray masses of conidia [22, 23] (**Figure 5**).

The presence of ethylene in storage can stimulate respiration rates and gray mold growth. Moreover, color of raspberries can be adversely affected by ethylene as it causes darkening of the red fruit to purple-red [10].

Figure 5. *Botrytis cinerea* growth on stored blackberries.

Another physiological disorder that can affect berries during storage is water loss, which in turn causes fruit shriveling, loss of gloss, and plays an important role in anthocyanin degradation. Water loss accelerates senescence of the fruit and the maximum permissible amount of water that can be lost (based on weight loss) from the fruit before becoming unmarketable is 6%. During postharvest handling of the berries, water loss can be reduced by prompt precooling and adequate packaging and storage at optimum temperature and relative humidity [24].

At present, the most extended methods used to maintain quality and bioactive compounds stability and to control decay of fruits and vegetables are postharvest washing, rapid cooling immediately after harvest, and storage at low temperatures [25, 26]. Furthermore, postharvest diseases are normally controlled by the use of synthetic fungicides [27] and storing under controlled or modified atmospheres with high CO_2 concentrations [28]. However, these methods present some limitations. Berries washing before retail is not recommended because the skin of the fruit may be damaged easily and the drying period delays precooling and enhances infections by pathogenic microorganisms [27].

Likewise, chemical fungicides may exert several negative effects on food safety and the environment, and there are public concerns about environmental pollution, possible contamination of berries by fungicide residues, and the inability to control fungal diseases because of the appearance of fungicide-tolerant strains of pathogens [29].

4.1. Temperature

One of the main factors affecting the storage shelf-life and quality of fruit and vegetables is temperature, as it regulates the rate of all the metabolic processes that occur in these products.

Low temperatures slow down fungal growth and, at the same time, reduce respiration rate and water loss and, therefore, delay ripening and senescence processes [30]. As these berries are insensitive to chilling injury, extending the shelf-life of berry fruit is often achieved through low temperature with optimum storage conditions for strawberries, raspberries, and blackberries being 0°C and 90–95% relative humidity [31].

Storage temperature is one of the key factors in suppressing fungal decay and influencing the stability of phenolic antioxidants in fruits during postharvest storage [32]. Also, temperature management is the most important factor to be taken into account to retain the initial ascorbic acid content during storage. Nevertheless, even when temperatures around 0°C are considered the best for berries' storage, the distribution in trucks and shops, the commercialization, and the storage in consumer households usually occur at higher temperatures, which can affect the berries shelf-life and their physicochemical quality and nutritional value, in terms of soluble sugars, vitamin C, and antioxidant compounds [33, 34].

Despite the already-known positive effects of low temperatures on postharvest shelf-life and quality of fresh fruits and vegetables, contradictory results can be found in the literature. Cooling of fruit at 0°C can be detrimental for short-term sales, as fruit appearance may be duller, and condensation of fruit during re-warming might result in greater decay incidence [35]. Jin et al. [36] indicated that strawberries stored at 10°C had higher antioxidant enzymes activities, higher level of phenolics and anthocyanins contents, and stronger oxygen radical scavenging capacities than those stored at 0 or 5°C, and Kalt et al. [32] found that low temperatures could affect anthocyanin synthesis during storage of small fruits. Similarly, anthocyanin and ascorbic acid biosynthesis was delayed in three strawberry cultivars stored at 6°C in comparison with storage at 16°C, while the contents of flavonols, ellagic acid, and total phenolics were not affected by the temperature lowering [33].

In blackberries, Joo et al. [37] found a reduction in total anthocyanin content (TAC) after 18 days at 3°C, while Wu et al. [38] did not see a clear tendency in the evolution of anthocyanins during 7 days at 2°C. On the contrary, in Andean blackberries harvested at the light and dark-red maturity stages, we observed an increase in total anthocyanin content during 10 days of storage at 8 ± 1°C and similar results were reported by Kim et al. [39] who observed that TAC increased after 15 days of storage at 1 or after 13 days at 1 plus 2 days at 20°C.

On the contrary, Piljac-Žegarac and Šamec [2] reported that the marketable quality of strawberries was preserved at 4°C for a prolonged period of time in comparison with storage at room temperature, while higher antioxidant capacity values were maintained at the lower temperatures, as opposed to 25°C. Similarly, storage of strawberries at 1°C together with moisture loss control reduced losses of total ascorbic acid by 7.5-fold compared to fruit stored at 20°C [40]. These authors concluded that even short periods at ambient temperature without control of water loss could result in considerable losses of total AA in strawberries. Moreover, Shin et al. [35] reported that the best temperature for visual appearance of strawberries was 0.5°C, but for short-term storage periods (up to 4 days), it was also possible to use moderate temperatures of 10°C. This temperature was useful to delay fruit ripening compared to room temperature and, at the same time, provided a balance between sensory attributes and those associated with the nutritional status of the fruit.

In Andean blackberries, we also observed that by storing the fruit in refrigerated storage, weight and firmness loss were reduced and microbial growth was delayed in comparison with storage at room temperature. What's more, the refrigerated fruit presented higher scores in the sensory analysis and, the total phenolic content and the antioxidant activity of the fruit were not affected by the cold storage.

4.2. Modified and controlled atmospheres

Modified atmosphere (MA) and controlled atmosphere (CA) refer to any atmosphere different from the normal air and usually involve atmospheres with reduced O_2 and/or elevated CO_2 levels. The difference between them is that CA is strictly controlled during all time.

Both MA and CA can be used for the storage, transport, and packaging of different types of food in compliment to low temperatures to extend their shelf-lives after harvest. Exposure of fresh horticultural crops to low O_2 and/or elevated CO_2 atmospheres within the range tolerated by each commodity reduces their respiration and ethylene production rates and therefore results in several beneficial effects such as delay of ripening and senescence and associated biochemical and physiological changes, reduction of sensitivity to ethylene action, alleviation of certain physiological disorders such as chilling injury, direct, and indirect control of pathogens, and consequently decay incidence and severity. On the contrary, if horticultural products are exposed to O_2 concentrations below, and/or CO_2 concentrations above their optimum tolerable range, the initiation and/or aggravation of certain physiological disorders, irregular ripening, increased susceptibility to decay, development of off-flavors, and eventually the loss of the product can occur [41].

Modified atmospheres (MA) and controlled atmospheres (CA) with elevated (15–20%) carbon dioxide and 5–10% oxygen concentrations reduce the growth of *Botrytis cinerea* (gray mold rot) and other decay-causing organisms. In addition, it reduces the respiration and softening rates of berries, thereby extending postharvest life. Nevertheless, further reductions of O_2 concentrations to 2 kPa had no benefit and may cause fermentation of the fruit [42].

In addition, off odors can be produced if the fruits are kept under high CO_2 atmospheres for more than 4 days as a result of anaerobic respiration [43] and the effect on the flavor preservation of these fruit is not clear. Several authors [28, 44, 45] reported changes in pH, titratable acidity, total soluble solids, sugars and organic acids, and fermentative metabolites after storage under CO_2-enriched atmospheres. In effect, different fermentative volatiles (acetaldehyde, ethanol, and ethyl acetate) were found after storage of strawberries in air + 20-kPa CO_2 at 2.8°C [46]. Among aroma compounds, esters are apparently the volatiles most affected by CO_2-enriched atmospheres [47].

Anthocyanin synthesis continues after harvest, but it is inhibited in fruits stored in high CO_2 concentrations. According to Holcroft and Kader [28], high CO_2 concentrations together with low O_2 concentrations can also affect adversely total ascorbic acid and anthocyanin contents and, thus, have a negative impact on fruit color and nutritional value. Conversely, the firmness, the external color, and the total phenolic compounds content of Selva strawberries were not affected by storage atmospheres with up to 20-kPa CO_2 [28]. In both, strawberries and raspberries, a cultivar-dependent response to changes in the storage atmosphere was observed.

An atmosphere of 12.5-kPa CO_2 and 7.5-kPa O_2 was effective in reducing decay in red raspberries and elevated concentrations of CO_2 together with reduced concentrations of O_2 were shown to inhibit mycelial growth and spore germination of *B. cinerea* and other fungi responsible for postharvest decay of fruit [42]. What's more, raspberries stored in 10/15-kPa O_2/CO_2 presented a more attractive color in comparison with fruit stored in air.

Finally, Giovanelli et al. [48] reported that the use of high and medium barrier materials delayed senescence and did not affect negatively the nutritional and antioxidant properties of red raspberries stored at 4°C. However, fermentative volatiles were found for these fruits, especially when an oxygen absorber was included in the packages.

5. Alternative methods for decontamination of strawberries

The need to minimize chemicals use has encouraged the rapid development of alternative techniques [29]. One of the new approaches is the use of 'generally recognized as safe' (GRAS) products, such as UV radiation and ozone, due to minimal concerns about their environmental impact and low residues in the treated commodity.

5.1. UV-C radiation

One strategy that can be an adjunct to refrigeration is the exposure of fruits to hormetic doses of UV-C radiation, a physical treatment that has been tested in strawberries and other fruits to control postharvest diseases [43, 49, 50] and delay some ripening-associated processes [51, 52]. Hormesis has been defined as the use of potentially harmful agents at low doses in order to induce a beneficial stress response [53].

UV-C seems to have a direct germicide effect on pathogens and an indirect effect by inducing defense mechanisms in the plant tissues [53–55]. Irradiation with UV-C is known to stimulate the phenylpropanoid pathway in several fruits, mainly by the induction of phenylalanine ammonia lyase (PAL), a key enzyme in this pathway [56]. The compounds synthesized by this pathway are implicated in a protective role against pathogens through reinforcement of plant cell walls, direct inhibition of growth, and/or inactivation of enzymes that contribute to tissue maceration [57].

Particularly in strawberry fruit, different UV-C doses increased enzyme activity, the antioxidant capacity and total phenolic content during storage, which correlated with lower fruit decay observed in treated fruit [58]. The synthesis and accumulation of phenolic compounds following irradiation with UV-C could also play an additional indirect role in fruit protection acting as natural substrates of polyphenol oxidase (PPO). One of the proposed roles of the reaction products of PPO (quinones) in plant defense is their action as bactericidals and fungicidals [59]. It was found that postharvest UV-C treatment, a few hours prior to inoculation with *B. cinerea*, reduced the percentage of fruit infection in strawberries during storage [60]. These authors also reported that irradiation of fruit with UV-C increased expression and activity of several enzymes (PAL, peroxidases, PPO, chitinases, and β-1,3-glucanases) which are involved in defense mechanisms against pathogens and abiotic stressors. In another experiment, photochemical treatment with UV-C delayed the appearance of gray mold rot in stored strawberries by up to 5 days at both 4 and 13°C [43]. These authors observed that the treatment with UV-C also enhanced the accumulation of anthocyanins, which in turn contributed to redder and visually more appealing fruit. In contrast, Erkan et al. [58] found little effect of UV-C treatments on anthocyanin content in strawberries.

5.2. Use of ozone

Another emerging technology with potential application in the food industry is the use of ozone as a sanitizer [61]. O_3 can be used for the postharvest treatment of fresh fruits and vegetables, in air or water, or as a continuous or intermittent atmosphere throughout the storage period. Gaseous ozone can be used to sanitize storage rooms and to prevent bacteria, molds, and yeasts development on the food surfaces. It can also eliminate undesirable flavors produced by bacteria and chemically remove ethylene gas to slow down the ripening process [62].

Ozone gas efficacy to inactivate microorganisms is conditioned by the species considered, its growth stage, the ions present in the air, the O_3 concentration and exposure time, and, the temperature and relative humidity of the room [63]. In air, the reactivity of ozone is greatest with fungi, molds, and some odor-causing chemicals and least with dry spores and bacteria. For optimum efficiency, it is also essential that the gas is thoroughly and evenly distributed quickly. Otherwise, decomposition will occur before the O_3 is able to contact its target [62].

Washing strawberries with ozonated water (0.3 ppm, 2 min) was an effective treatment to reduce microbial counts and enhance anthocyanin and ascorbic acid retention of these fruits during 13 days of refrigerated storage [64]. Similarly, Zhang et al. [65] reported greater ascorbic acid retention in strawberries treated with gaseous O_3 (4 ppm, 30 min/day) in comparison with untreated fruit. Moreover, strawberries' levels of biothiols were not affected by the treatment with either gaseous-phase or aqueous-phase ozone [66]. Finally, while total phenolic and ellagitanin contents were similar in O_3-treated and untreated strawberry fruit after a storage period of 12 days, the procyanidins and anthocyanins contents were reduced by the exposure to this gas [67].

In addition to their antimicrobial power, O_3 and UV-C radiation gather other advantages, which turn them into appealing and environmental friendly technologies [61]. Neither ozone nor UV-C leave undesirable residues on food or food-contact surfaces nor create undesirable disinfection by-products [68]. Moreover, the application of these sanitizers in food processing is approved by the code of Food and Drug Administration (FDA) in the USA and is allowed by organic certification [69].

There are numerous studies in the literature reporting on the use of both, O_3 and UV-C light, on several fruits and vegetables. However, results are sometimes contradictory and information about the effects of these decontamination treatments on sensory and nutritional quality or health-promoting composition of treated products is scarce. It should be taken into account that while high doses of oxidizing agents may result in depletion of natural antioxidants, moderate or low doses of oxidative stress were shown to cause a protective response, enhancing the level of endogenous antioxidants [70, 71].

6. Transport

Berries must be transported in clean and well-maintained trucks, and it is crucial to maintain the fruit cold and wrapped during loading, unloading, and transportation. In order to ensure a proper circulation of the cold air, the flats or boxes must be stacked on pallets and without touching the truck walls. Frequently, rural roads are not in optimum conditions causing bruises and abrasion due to the truck vibration and the impacts between the packed fruit and between the fruit and the walls of the packs. These mechanical damages can be minimized by stabilizing the load on every pallet, for example, by using stretch film and by using trucks with air suspension systems.

When refrigerated transport is used, it must be considered that trucks' mechanical refrigeration equipment is designed to maintain temperature but they do not have the capacity to lower the temperature of the produce. So, it is very important to achieve the proper cooling of the product before loading. Finally, to avoid condensation on the fruits, berries should be allowed to warm only when they are ready for display to consumers and before removing the plastic wrap over the flats [10].

Author details

Sandra Horvitz

Address all correspondence to: sandra.horvitz@unavarra.es

Food Science and Engineering Faculty, Technical University of Ambato, Ambato, Ecuador

References

[1] Basu A, Nguyen A, Betts NM, Lyons TJ. Strawberry as a functional food: An evidence-based review. Critical Reviews in Food Science. 2014;**54**(6):790-806

[2] Piljac-Žegarac J, Šamec D. Antioxidant stability of small fruits in postharvest storage at room and refrigerator temperatures. Food Research International. 2011;**44**:345-350

[3] Sánchez MT, Haba MJDL, Benítez-López M, Fernández-Novales J, Garrido-Varo A, Pérez-Marín D. Non-destructive characterization and quality control of intact strawberries based on NIR spectral data. Journal of Food Engineering. 2012;**110**:102-108

[4] Kalt W, Prange RK, Lidster PD. Postharvest color development of strawberries: Influence of maturity, temperature and light. Canadian Journal of Plant Science. 1993;**73**:541-548

[5] Sturm K, Koron D, Stampar F. The composition of fruit of different strawberry varieties depending on maturity stage. Food Chemistry. 2003;**83**(3):417-422

[6] UNECE (United Nations Economic Commission for Europe). UNECE STANDARD FFV-57 Concerning the Marketing and Commercial Quality Control of Berry Fruits; United Nations. New York and Geneva. 2011

[7] Krüger E, Schöpplein E, Rasim S, Cocca G, Fischer H. Effects of ripening stage and storage time on quality parameters of red raspberry fruit. European Journal of Horticultural Science. 2003;**68**(4):176-182

[8] García M. La agroindustria de la mora. Alternativas viables para los fruticultores. Tecnología para el Agro. 2001;**1**(2):15-17

[9] Rivera A, Tong CB. Commercial Postharvest Handling of Strawberries (*Fragaria* spp.) [Internet]. 2013. Available from: http://www.extension.umn.edu/garden/fruit-vegetable/commercial-postharvest-handling-of-strawberries/ [Accessed: February 15, 2017]

[10] Bushway L, Pritts M, Handley D. Raspberry and Blackberry Production Guide for the Northeast, Midwest, and Eastern Canada, NRAES-35. Natural Resource, Agriculture, and Engineering Service (NRAES); Ithaca, New York, USA. 2008. p. 158

[11] Stavang JA, Freitag S, Foito A, Verrall S, Heide OM, Stewart D, et al. Raspberry fruit quality changes during ripening and storage assessed by colour, sensory evaluation and chemical analyses. Scientia Horticulturae. 2015;**195**:216-225

[12] Forney CF, Kalt W, McDonald JE, Jordan MA. Changes in strawberry fruit quality during ripening and off the plant. Acta Horticulturae. 1998;**464**:506

[13] Miszczak A, Forney CF, Prange RK. Development of aroma volatiles and color during postharvest ripening of 'Kent' strawberries. Journal of the American Society for Horticultural Science. 1995;**120**:650-655

[14] Sacks EJ, Shaw DV. Color change in fresh strawberry fruit of seven genotypes stored at 0°C. HortScience. 1993;**28**:209-210

[15] Azodanlou R, Darbellay C, Luisier JL, Villettaz JC, Amado R. Quality assessment of strawberries (*Fragaria* species). Journal of Agricultural and Food Chemistry. 2003;**51**:715-721

[16] Kader AA. Standardization and inspection of fresh fruits and vegetables. In: Kader AA, editor. Postharvest Technology of Horticultural Crops. 3rd ed. Oakland, CA: University of California, Division of Agriculture and Natural Resources; 2002. pp. 287-360

[17] Instituto Ecuatoriano de Normalización (INEN). INEN 2427: Frutas frescas. Mora. Requisitos; 2010

[18] Yang FM, Li HM, Li F, Xin ZH, Zhao LY, Zheng YH, et al. Effect of nano-packing on preservation quality of fresh strawberry (*Fragaria ananassa* Duch. cv Fengxiang) during storage at 4°C. Journal of Food Science. 2010;**75**(3):C236-C240

[19] Storage of Berries [Internet]. 2017. Available from: http://www.omafra.gov.on.ca/english/crops/facts/storage_berries.htm [Accessed: January 20, 2017]

[20] Goulas V, Manganaris GA. The effect of postharvest ripening on strawberry bioactive composition and antioxidant potential. Journal of the Science of Food and Agriculture. 2011;**91**:1907-1914

[21] Jing W, Tu K, Shao XF, Su ZP, Zhao Y, Wang S, et al. Effect of postharvest short hot-water rinsing and brushing treatment on decay and quality of strawberry fruit. Journal of Food Quality. 2010;**33**:262-272

[22] Williamson B, Tudzynski B, Tudzynski P, Van Kan JAL. *Botrytis cinerea*: The cause of grey mould disease. Molecular Plant Pathology. 2007;**8**:561-580

[23] Hassenberg K, Geyer M, Ammon C, Herppich WB. Physico-chemical and sensory evaluation of strawberries after acetic acid vapour treatment. European Journal of Horticultural Science. 2011;**76**(4):125-131

[24] Nunes MCN, Brecht JK, Morais AMMB, Sargent SA. Possible influences of water loss and polyphenol oxidase activity on anthocyanin content and discoloration in fresh ripe strawberry (cv. Oso Grande) during storage at 1°C. Journal of Food Science. 2005;**70**(1):S79–S84

[25] Han C, Zhao Y, Leonard SW, Trabe MG. Edible coatings to improve storability and enhance nutritional value of fresh and frozen strawberries (*Fragaria × ananassa*) and raspberries (*Rubus ideaus*). Postharvest Biology and Technology. 2004;**33**(1):67-78

[26] Hernández-Muñoz P, Almenar E, Ocio MJ, Gavara R. Effect of calcium dips and chitosan coatings on postharvest life of strawberries (*Fragaria × ananassa*). Postharvest Biology and Technology. 2006;**39**(3):247-253

[27] Vardar C, Ilhan K, Karabulut OA. The application of various disinfectants by fogging for decreasing postharvest diseases of strawberry. Postharvest Biology and Technology. 2012;**66**:30-34

[28] Holcroft DM, Kader AA. Controlled atmosphere induced changes in pH and organic acid metabolism may affect color of stored strawberry fruit. Postharvest Biology and Technology. 1999;**17**:19-32

[29] Leroux P. Chemical control of botrytis and its resistance to chemical fungicides. In: Elad Y, Williamson B, Tudzynski P, Delen N, editors. Botrytis: Biology, Pathology and Control. Dordrecht: Springer; 2007. pp. 195-222

[30] Oliveira DM, Rosa CILF, Kwiatkowski A, Clemente E. Biodegradable coatings on the postharvest of blackberry stored under refrigeration. Revista Ciencia Agronómica. 2013;**44**:302-309

[31] Giuggioli NR, Briano R, Baudino C, Peano C. Effects of packaging and storage conditions on quality and volatile compounds of raspberry fruits. CyTA—Journal of Food. 2015;**13**(4):512-521

[32] Kalt W, Forney CH, Martin A, Prior RL. Antioxidant capacity, vitamin C, phenolics, and anthocyanins after fresh storage of small fruits. Journal of Agricultural and Food Chemistry. 1999;**47**:4638-4644

[33] Cordenunsi BR, Genovese MI, Oliveira do Nascimento JR, Hassimotto NMA, dos Santos RJ, Lajolo FM. Effects of temperature on the chemical composition and antioxidant activity of three strawberry cultivars. Food Chemistry. 2005;**91**:113-121

[34] Krüger E, Dietrich H, Schöpplein E, Rasim S, Kürbel P. Cultivar, storage conditions and ripening effects on physical and chemical qualities of red raspberry fruit. Postharvest Biology and Technology. 2011;**60**:31-37

[35] Shin Y, Liu RH, Nock JF, Holliday D, Watkins CB. Temperature and relative humidity effects on quality, total ascorbic acid, phenolics and flavonoid concentrations, and antioxidant activity of strawberry. Postharvest Biology and Technology. 2007;**45**:349-357

[36] Jin P, Wang SY, Wang CY, Zheng Y. Effect of cultural system and storage temperature on antioxidant capacity and phenolic compounds in strawberries. Food Chemistry. 2011;**124**:262-270

[37] Joo M, Lewandowski N, Auras R, Harte J, Almenar E. Comparative shelf life study of blackberry fruit in bio-based and petroleum-based containers under retail storage conditions. Food Chemistry. 2011;**126**(4):1734-1740

[38] Wu R, Frei B, Kennedy JA, Zhao Y. Effects of refrigerated storage and processing technologies on the bioactive compounds and antioxidant capacities of 'Marion' and 'Evergreen' blackberries. LWT-Food Science and Technology. 2010;**43**:1253-1264

[39] Kim MJ, Perkins-Veazie P, Ma G, Fernandez G. Shelf life and changes in phenolic compounds of organically grown blackberries during refrigerated storage. Postharvest Biology and Technology. 2015;**110**:257-263

[40] Nunes MCN, Brecht JK, Morais AMMB, Sargent SA. Controlling temperature and water loss to maintain ascorbic acid levels in strawberries during postharvest handling. Journal of Food Science. 1998;**63**(6):1033-1036

[41] Yahia EM. Modified and controlled atmospheres for the storage, transportation, and packaging of horticultural commodities. In: Yahia EM, editor. Modified and Controlled Atmospheres for the Storage, Transportation, and Packaging of Horticultural Commodities. Boca Raton, FL: CRC Press; 2009. pp. 1-16

[42] Forney CF, Jamieson AR, Pennell KDM, Jordan MA, Fillmore SAE. Relationships between fruit composition and storage life in air or controlled atmosphere of red raspberry. Postharvest Biology and Technology. 2015;**110**:121-130

[43] Baka M, Mercier J, Corcuff R, Castaigne F, Arul J. Photochemical treatment to improve storability of fresh strawberries. Journal of Food Science. 1999;**64**(6):1068-1072

[44] Sanz C, Perez AG, Olias R, Olias JM. Quality of strawberries packed with perforated polypropylene. Journal of Food Science. 1999;**64**:748-752

[45] Gil MI, Holcroft DM, Kader AA. Changes in strawberry anthocyanins and other polyphenols in response to carbon dioxide treatments. Journal of Agricultural and Food Chemistry. 1997;**45**:1662-1667

[46] Watkins CB, Manzano-Mendez JE, Nock JF, Zhang J, Maloney KE. Cultivar variation in response of strawberry fruit to high carbon dioxide treatments. Journal of the Science of Food and Agriculture. 1999;**79**:886-890

[47] Pelayo-Zaldívar C, Abda JB, Ebeler SE, Kader AA. Quality and chemical changes associated with flavor of Camarosa strawberries in response to a CO_2-enriched atmosphere. HortScience. 2007;**42**(2):299-303

[48] Giovanelli G, Limbo S, Buratti S. Effects of new packaging solutions on physico-chemical, nutritional and aromatic characteristics of red raspberries (*Rubus idaeus* L.) in postharvest storage. Postharvest Biology and Technology. 2014;**98**:72-81

[49] Marquenie D, Michiels CW, Impe JFV, Schrevens E, Nicolai BN. Pulsed white light in combination with UV-C and heat to reduce storage rot of strawberry. Postharvest Biology and Technology. 2003;**28**:455-461

[50] Perkins-Veazie P, Collins JK, Howard L. Blueberry fruit response to postharvest application of ultraviolet radiation. Postharvest Biology and Technology. 2008;**47**:280-285

[51] Pan J, Vicente AR, Martínez GA, Chaves AR, Civello PM. Combined use of UV-C irradiation and heat treatment to improve postharvest life of strawberry fruit. Journal of the Science of Food and Agriculture. 2004;**84**:1831-1838

[52] Pombo MA, Dotto MC, Martínez GA, Civello PM. UV-C irradiation delays strawberry fruit softening and modifies the expression of genes involved in cell wall degradation. Postharvest Biology and Technology. 2009;**51**:141-148

[53] Shama G, Alderson P. UV hormesis in fruits: A concept ripe for commercialization. Trends in Food Science & Technology. 2005;**16**:128-136

[54] Huyskens-Keil S, Hassenberg K, Herppich WB. Impact of postharvest UV-C and ozone treatment on textural properties of white asparagus (*Asparagus officinalis* L.). Journal of Applied Botany and Food Quality. 2011;**84**:229-234

[55] Civello PM, Vicente AR, Martínez GA. UV-C technology to control postharvest diseases of fruits and vegetables. In: Recent Advances in Alternative Postharvest Technologies to Control Fungal Diseases in Fruit and Vegetables. Transworld Research Network; Kerala, India, 2007. pp. 71-207

[56] Eichholz I, Rohn S, Gamm A, Beesk N, Herppich WB, Kroh LW, et al. UV-B mediated flavonoid synthesis in white asparagus (*Asparagus officinalis* L.). Food Research International. 2012;**48**:196-201

[57] Treutter D. Significance of flavonoids in plant resistance and enhancement of their biosynthesis. Plant Biology. 2005;**7**:581-591

[58] Erkan M, Wang SY, Wang CY. Effect of UV treatment on antioxidant capacity, antioxidant enzyme activity and decay in strawberry fruit. Postharvest Biology and Technology. 2008;**48**:163-171

[59] Yoruk R, Marshall MR. Physicochemical properties and function of plant polyphenol oxidase: A review. Journal of Food Biochemistry. 2003;**27**(5):361-422

[60] Pombo MA, Rosli HG, Martínez GA, Civello PM. UV-C treatment affects the expression and activity of defense genes in strawberry fruit (*Fragaria x ananassa*, Duch.). Postharvest Biology and Technology. 2011;**59**:94-102

[61] Alexandre EMC, Santos-Pedro DM, Brandão TRS, Silva CLM. Influence of aqueous ozone, blanching and combined treatments on microbial load of red bell peppers, strawberries and watercress. Journal of Food Engineering. 2011;**105**:277-282

[62] Rice RG, Farquhar JW, Bollyky LJ. Review of the applications of ozone for increasing storage times of perishables foods. Ozone: Science & Engineering. 1982;**4**(3):147-163

[63] Pascual A, Llorca I, Canut A. Use of ozone in food industries for reducing the environmental impact of cleaning and disinfection activities. Trends in Food Science & Technology. 2007;**18**:S29-S35

[64] Alexandre EMC, Brandão TRS, Silva CLM. Efficacy on non-thermal technologies and sanitizer solutions on microbial load reduction and quality retention of strawberries. Journal of Food Engineering. 2012;**108**:417-426

[65] Zhang X, Zhang Z, Wang L, Zhang Z, Li J, Zhao C. Impact of ozone on quality of strawberry during cold storage. Frontiers of Agriculture in China. 2011;**5**(3):356-360

[66] Demirkol O, Cagri-Mehmetoglu A, Qiang Z, Ercal N, Adams C. Impact of food disinfection on beneficial biothiol contents in strawberry. Journal of Agricultural and Food Chemistry. 2008;**56**:10414-10421

[67] Horvitz S, Cantalejo MJ. Application of ozone for the postharvest treatment of fruits and vegetables. Critical Reviews in Food Science. 2014;**54**(3):312-339

[68] Zhang L, Lu Z, Yu Z, Gao X. Preservation of fresh-cut celery by treatment of ozonated water. Food Control. 2005;**16**(3):279-283

[69] Horvitz S, Cantalejo MJ. Effects of gaseous O_3 and modified atmosphere packaging on the quality and shelf-life of partially dehydrated ready-to-eat pepper strips. Food and Bioprocess Technology. 2015;**8**(8):1800-1810

[70] Cisneros-Zevallos L. The use of controlled post-harvest abiotic stresses as a tool for enhancing the nutraceutical content and adding value to fresh fruits and vegetables. Journal of Food Science. 2003;**68**:1560-1565

[71] Rodov V, Vinokur Y, Horev B, Goldman G, Moroz A, Shapiro A. Phobiological treatment: A way to enhance the health value of fruits and vegetables? In: The Use of UV as a Postharvest Treatment: Status and Prospects. Proceeding of the COST Action 924 Work Group Meeting. Antalya; 2006. pp. 64-70

4

Modified Atmosphere Packaging: Design and Optimization Strategies for Fresh Produce

Diego A. Castellanos and Aníbal O. Herrera

Additional information is available at the end of the chapter

Abstract

Modified atmosphere packaging (MAP) is a useful preservation system that allows to significantly increase the shelf life of fruits and vegetables. The MAP results of changing the composition of the atmosphere in the packaging headspace due to the dynamic interaction between the metabolic processes of the packaged product on the one hand, in which O_2 is consumed and another gases such as CO_2 and water vapor are generated, and on the other hand, by transferring all of these gases through the package. The aim of the system is to balance these two processes in such a way that constant levels of these different gases are reached in the packaging headspace and that these equilibrium levels are as favorable as possible to preserve the product. This chapter describes design strategies to obtain a satisfactory gas transfer capacity in the MAP system through the configuration of several related variables such as the type of packing material, its thickness, the transfer surface area and the required number and diameter of perforations. For this, the necessary steps are proposed to estimate the appropriate transfer capacity according to the equilibrium gas concentrations desired to longer preserve the product by using the mass balance equations of the MAP system.

Keywords: packaging configuration, gas transfer capacity, perforation, preservation, fruits, vegetables

1. Introduction

After harvest, the fresh produce undergoes a series of biochemical transformations that lead to the development of color, aroma and characteristic flavors, reduction of acidity and tissue softening [1, 2]. These processes continue until the substrates that support them are consumed, after which follows the deterioration or senescence phase. The postharvest period involves a series of biochemical reactions or metabolic activities that can lead to an increase in the concentration of organic acids, lipids, phenolic compounds, generation of volatile compounds (aroma), variations in the activity of enzymes, degradation of chlorophyll and biosynthesis of pigments, degradation of pectins and conversion of starch in sugars, which will cause loss of firmness and weight, and increase in the sweetness perception [1, 3–5]. In the case of edible products, all previous processes lead to the development of characteristics that make them acceptable for consumption [1, 6]. In this period, the plant cells continue their respiratory activity, using oxygen from the surrounding atmosphere and releasing carbon dioxide as result of a series of oxide-reduction reactions of the substrates present [7, 8].

The respiratory response of fresh produce in postharvest and the development of ripening and senescence processes depends on the temperature, of the atmosphere surrounding the stored product and also on the levels of gases such as O_2, CO_2, ethylene and water vapor, which strongly influence its metabolism. For this reason, by regulating the temperature and gas concentrations in the atmosphere surrounding the product, it is possible to reduce the development of its different biochemical processes and increase its preservation time [9–11]. Likewise, the growth of microorganisms that may cause deterioration or other potentially pathogenic organisms will also be closely related to environmental conditions due to their development, and activity can be controlled by modifying these conditions [9].

1.1. Modified atmosphere packaging

Modified atmosphere packaging (MAP) has been constituted as a preservation method that allows to significantly increase the shelf life of fruits and vegetables. The MAP results of changing the composition of the atmosphere surrounding the product in order to reduce its natural deterioration and the microbial spoilage [9, 12]. In a MAP system, there is a change in the concentration of gases in the packaging headspace due to the dynamic interaction between the metabolic and biochemical processes of the packaged product on the one hand, in which O_2 is consumed and CO_2, ethylene and water vapor are generated, and on the other hand, the transfer of all of these gases through the packaging. Accordingly, O_2 from the external atmosphere will be entering through the packaging to replace that the product is consuming and the other gases will be leaving out through the packaging system as surplus [11, 13–15]. The aim of the system is to balance these two processes in such a way that constant levels of these different gases are reached in the packaging headspace and that these equilibrium levels are as favorable as possible to preserve the product [9].

In the first MAP systems, a reference was made to active and passive atmospheres depending on whether a volume of gas with a different atmospheric concentration was introduced at the time of sealing the package (active MAP), or simply the bag was sealed with atmospheric air (passive MAP). This is of limited utility if it is not considered that regardless of the initial concentration of gases in the packaging headspace, the final concentration of these will depend on the interaction between the metabolic processes of the product (and possible microorganisms present) and the exchange of gases through the packaging. Failure to know the different rates at which these processes are carried out will result in an inadequate MAP, resulting in concentrations of gases that do not contribute to preserving the product at the best or that directly increase its deterioration at the worst [9, 14]. For this reason and in order to obtain a satisfactory MAP, it is necessary to design the packaging system previously, and this basically consists of determining the gas transfer capacity in the packaging that is required to balance the metabolic processes of the product to be packed and in this way, to achieve favorable gas concentrations for its preservation [13].

After packing a fruit or vegetable, the O_2 concentration in the headspace will decrease, and the CO_2 concentration will increase until reaching constant values once the equilibrium is reached. In these conditions of reduced O_2 and high CO_2, the metabolic processes of the product and the microbial activity slow down increasing its shelf life. Likewise, as the storage temperature decreases, there is also a reduction in the development of these processes resulting in an additional increase of the produce shelf life [11]. On the other hand, low O_2 levels, high CO_2 levels and above all, low temperatures in the packaging system can significantly decrease the growth of microorganisms that cause deterioration in the product and of other ones potentially pathogenic [16]. In these unfavorable circumstances for microbial growth (low O_2, high CO_2 and low temperature), a competition is established between the produce, natural microflora and other external and potentially pathogenic microorganisms that has an inhibitory effect for the development of the latter ones [17]. As shown in **Table 1**, for a wide range of fresh produce, the most favorable levels in the packaging headspace for preservation are between 1 and 10% of O_2, 5 and 15% of CO_2 and 80 and 90% RH, with temperatures as low as possible (0–15°C) but without inducing chilling injury [9, 18, 19].

1.2. MAP uses to preserve vegetables and fruits

MAP has been successfully applied to preserve several vegetables including broccoli, cauliflower, carrots, garlic and others [20]. The effect of the packaging has been studied on the quality of broccoli [21] and cauliflowers [22]. In order to preserve these products, micro or macro perforated polyethylene and polypropylene bags have been used to compensate for the respiration and transpiration of the packed product and to achieve favorable equilibrium gas levels for its preservation [11, 23]. Other vegetables have been successfully preserved in MAP systems, including cabbage [24], cucumber [25], onion [26], spinach [27] and tomato [28]. Refrigeration temperatures were used in all cases, although at room temperatures, the packaging system may continue to have a preservation effect, although somewhat lower [1, 9].

As for fruits, MAP systems have been used successfully to preserve apple [29], banana [30], breba [31], blueberry [32], feijoa [18], grape [33], guava [34], mandarin [10], medlar [15], papaya [35], pear [36], pomegranate [37], strawberry [38] and others. In all these cases, it was possible to increase the product shelf life between 40% and up to twice with respect to atmospheric conditions. Recommended O_2 and CO_2 concentrations in the MAP system are shown in **Table 1** for various fruits and vegetables.

1.3. Packaging materials used in MAP

Most MAP systems consist of flexible or semiflexible structures, bags of different sizes and thicknesses with and without perforations [39]. The most widely used polymer films for MAP

Produce	O_2 (%)	CO_2 (%)
Fruits		
Apple	2–3	1–3
Avocado	2–5	3–10
Banana	2–5	2–8
Feijoa	5–11	7–12
Mango	3–7	5–8
Orange	5–10	5–8
Pear	2–3	2–5
Strawberry	5–10	5–15
Vegetables		
Broccoli	2–3	5–10
Carrot	2–5	2–5
Lettuce	2–3	1–3
Onion	2–5	2–10
Pepper	3–5	1–3
Radish	2–5	2–5
Spinach	15–20	10–20
Tomato	3–5	5–11

Data compiled from [2, 11, 12, 18].

Table 1. Suitable O_2 and CO_2 concentrations in MAP for different fruits and vegetables.

are synthetic polyolefins, low-density polyethylene (LDPE), linear low-density polyethylene (LLDPE), high-density polyethylene (HDPE), polypropylene (PP), polyvinyl chloride (PVC), polyesters, polyethylene terephthalate (PET), polyvinylidene chloride (PVDC), ethylene-vinyl alcohol (EVOH), polyamide (Nylon), polyvinyl alcohol (PVOH), ethylene vinyl acetate (EVA), cellulose-derived plastics such as cellophane and natural biodegradable polymers like polylactic acid (PLA) [11, 19]. Each of these polymeric materials offers different mechanical characteristics and fundamentally different permeability to O_2, CO_2 and water vapor. Depending on the specific requirements of gas permeation, materials of very low permeability to O_2 and CO_2 such as polyamide, of moderate permeability such as polypropylene and EVA, or of high permeability as LDPE can be selected or even combinations of these materials can be made in several layers to adjust the MAP system to the produce respiration and transpiration [9, 39]. In addition, in order to avoid the accumulation of moisture inside the package while maintaining a reduced permeability to O_2 and CO_2, it is possible to use PLA or EVOH, which have a high permeability to water vapor [39, 40]. In **Table 2**, permeability (Q) coefficients to O_2, CO_2 and water vapor are for different polymeric materials.

Polymeric material	Q_{O_2}	Q_{CO_2}	Q_{N_2}	Q_{H_2O}*
Low-density Polyethylene (LDPE)	190–200	1050–1250	100–150	5500–6000
High-density Polyethylene (HDPE)	40–70	160–190	14–20	2100–2500
Cast Polypropylene (PP)	80–95	250–280	17–25	4000–4200
Oriented Polypropylene (OPP)	40–50	180–200	10–15	1800–2200
Polylactic Acid (PLA)	60–80	150–190	20–30	300000–350000
Ethylene vinyl acetate (EVA)	300–330	1100–1300	120–150	10000–15000
Polyamide (PA)	1–5	3–10	0.3–1	20000–30000

* Permeability coefficient to water vapor
Estimated from [9, 11, 13, 39, 40, 64]

Table 2. Permeability coefficients (cm^3 mm m^{-2} atm^{-1} d^{-1}) at 20°C for different packaging materials.

2. Dynamics of a MAP system

In the design of MAP systems for fresh produce, it is necessary to understand the dynamics of the interactions established between the product, the atmosphere generated in the packaging headspace and the packaging system itself. Each MAP design must be optimized for a specific product, since agricultural products have different metabolisms to each other and, as mentioned above, the MAP system must balance the processes of respiration, transpiration and gas permeation through the packaging that will be occurring simultaneously. Factors that affect both the metabolism of the packaged product and the gas permeation through the packaging should be considered for the design of the MAP system [9, 14, 41]. Among these factors are the type of produce and ripening stage, the characteristics of the packaging system and the storage conditions [10, 13, 42]. With regard to the packaged product, it is also necessary to know in advance its transpiration, oxygen consumption and CO_2 and ethylene generation rates at the packaging conditions and the optimum concentrations of these gases and relative humidity that favor the increase of its shelf life [10, 41, 43].

For the proper design of the MAP system, different mathematical models have been developed to represent the processes of respiration, transpiration and ethylene production for a variety of horticultural products [14, 44, 45]. In these models, it is intended to represent the rates of O_2 consumption and CO_2, ethylene and water vapor generation from the product as a function of the storage temperature, the concentration of these gases in the packaging headspace and the weight of the packaged product. In some studies, these metabolic processes developed in the fresh produce are described by using enzymatic kinetics equations [37, 46, 47], while in other studies, empirical or semiempirical equations are used to represent them [10, 48, 49]. As for the transfer of gases through the packaging and the perforations that it may have, generally have been developed models derived from the Fick's law, stating the mass balances for each gas in the system and considering the influence of the temperature at the gas permeation rates [14, 50].

2.1. Mathematical representation of the MAP system

The change in the composition of each gas in the packaging headspace will depend on the metabolic processes of the packaged product and the gas transfer through the packaging system.

$$
\begin{bmatrix}
Accumulation\ or \\
decrease\ rate\ of\ gas\ in \\
the\ packaging\ headspace
\end{bmatrix}
=
\begin{bmatrix}
Transfer\ rate\ of \\
gas\ through\ the \\
packaging\ system
\end{bmatrix}
+
\begin{bmatrix}
Generation\ or \\
consumption\ rate\ of \\
gas\ from\ the\ product
\end{bmatrix}
\tag{1}
$$

As mentioned above, when the product is packaged will begin to consume O_2 from the packaging headspace and to generate CO_2 due its respiration, water vapor due its transpiration and ethylene. This will create a concentration differential between the inside and outside of the package resulting in an O_2 inlet and an outlet of the other gases to the external atmosphere. The composition of the different gases will be changing until a balance is reached between the metabolic processes in the product and the transfer of gases through the package (Eq. (2)). At this point, the MAP system becomes an equilibrium modified atmosphere packaging (EMAP), which is a kind of self-sustaining controlled atmosphere system that will be maintained until the product substrates are consumed or until storage conditions such as temperature are altered.

$$
\begin{bmatrix}
Generation\ or \\
consumption\ rate\ of \\
gas\ from\ the\ product
\end{bmatrix}
=
\begin{bmatrix}
Transfer\ rate\ of \\
gas\ through\ the \\
packaging\ system
\end{bmatrix}
\tag{2}
$$

2.1.1. Respiration rate

Fresh produce shelf life depends on its respiration rate. Fruits such as mango and banana that have high respiration rates are highly perishable [7]. The energy released during respiration is associated with the energy required to completely oxidize a mole of substrate to CO_2 and H_2O. In the case of a mole of hexose that is the most common carbohydrate, approximately 2880 kJ are released. Each hexose molecule is oxidized to 6 CO_2 molecules, using 32 molecules of adenosine diphosphate (ADP) and six molecules of O_2 to form 12 water molecules (Eq. (3)). Under normal physiological conditions, 50 to 60% of this energy is chemically captured to form 32 molecules of adenosine triphosphate (ATP), which are required for subsequent cell metabolic processes [12, 45, 51]. As respiration is a series of oxidation reactions, the rate at which it is carried out will be related to the O_2 concentration in the medium, and when this concentration decreases, the energy released will also be reduced [52].

$$
C_6H_{12}O_6(s) + 6O_2 \rightarrow 6CO_2 + 6H_2O + energy
\tag{3}
$$

In the case of carbohydrates such as glucose or fructose, when these are completely oxidized, the O_2 volume used per unit time will be practically the same as the volume of generated CO_2 per unit time [12, 45]. For organic acid oxidation, more CO_2 volume is produced than the used O_2 and for lipids oxidation, less CO_2 volume is generated than O_2 used. From the above, the respiratory quotient (RQ) is defined as the ratio of CO_2 produced to O_2 consumed per unit time [53, 54]. The RQ values range between 0.7 and 1.3 for the majority of fresh produce [12]. High RQ values usually indicate anaerobic respiration in produce tissues or microbial activity that produces aldehydes and alcohols through fermentation. In such cases, a rapid change in the RQ can be used as an indication of the shift from aerobic to anaerobic respiration [2]. **Table 3** shows estimated O_2 consumption, CO_2 production rates and RQ values for different fruits and vegetables.

Produce	r_{O_2}	r_{CO_2}	RQ
Apple	340–380	350–450	1.02–1.20
Avocado	3500–3800	2400–2700	0.70–0.72
Banana	3400–3600	2600–2700	0.75–0.76
Broccoli	2200–2400	2000–2150	0.88–0.90
Feijoa	4500–4800	3600–3800	0.79–0.80
Guava	850–1000	800–950	0.95–0.98
Strawberry	1050–1150	1000–1100	0.94–0.96
Tomato	640–680	700–730	1.05–1.07

Estimated from [12–14, 41, 47, 54, 57]

Table 3. O_2 consumption and CO_2 production rates (cm^3 kg^{-1} d^{-1}) and respiratory quotient at 20°C and atmospheric air for different fresh produce.

Many mathematical models and equations have been used to represent the respiration process in fresh produce and the effect of the O_2 and CO_2 levels and storage temperature. These models include empirical equations, chemical kinetics and enzyme kinetics equations based on the Michaelis-Menten (MM) model [14, 42, 45, 47, 54–57]. Among these equations stand out the Michaelis-Menten kinetics that have been successfully used for many products, specifically the equation of uncompetitive inhibition (MMU) wherein the inhibitor (CO_2 in this case) reacts with the enzyme-substrate complex, reducing the overall respiration rates at high concentrations. Nevertheless, the maximum O_2 consumption or CO_2 production rates are not influenced by the CO_2 levels in the atmosphere surrounding the product [14, 45, 55, 58–60].

In the MMU kinetics, the rates of O_2 consumption and CO_2 generation are, respectively:

$$r_{O_2} = \frac{r_{O_{2max}} y_{O_2}}{K_{m_{O_2}} + y_{O_2}\left(1 + \frac{y_{CO_2}}{K_{mu_{CO_2}}}\right)} \tag{4}$$

$$r_{CO_2} = \frac{r_{CO_{2max}} y_{O_2}}{K_{m_{CO_2}} + y_{O_2}\left(1 + \frac{y_{CO_2}}{K_{mu'_{CO_2}}}\right)} \tag{5}$$

where r_{O_2} and r_{CO_2} are the O_2 consumption and CO_2 production rates, respectively (in cm^3 kg^{-1} d^{-1}); r_{O_2max} and r_{CO_2max} are the maximum rates of O_2 consumption or CO_2 production,

K_{mO_2} and K_{mCO_2} are the dissociation constants of the enzyme-substrate complex, and K_{muCO2} and $K_{mu'CO_2}$ are the inhibition constants due to CO_2 [14, 45, 47]. In other cases, competitive, non-competitive or combined MM models have been used to represent the respiration rates [45, 12].

Respiration rates are also directly influenced by temperature. As temperature increases, O_2 consumption and CO_2 production becomes higher, and this relationship has been satisfactorily described by considering each of the parameters of the Michaelis-Menten kinetics dependent on temperature by using the Arrhenius' Law [14, 47].

2.1.2. Transpiration rate

The transpiration rate can be defined as the water lost or evaporated in the product per unit time. The transpiration and water loss can be considered as a consequence of the heat transferred to and from the product and the vapor pressure difference between the product and the surrounding atmosphere [41, 46, 61, 62]. When the vapor pressure in the atmosphere is less than the vapor pressure in the product, it could be considered that both processes will contribute to transpiration [44]. When the vapor pressures become equal, transpiration only will be seen as a consequence of the heat exchange to and from the product. This heat will be transferred to evaporate the product moisture mainly from the fraction of dissipated energy in the respiration process (that than is not used in the product metabolism), and from the temperature difference between the product and the surrounding atmosphere in the process of cooling [41, 46]. Some part of this energy transferred will change the product temperature as sensible heat, while another part will be converted to latent heat through moisture evaporation. If the system is in thermal equilibrium, as in a temperature-controlled chamber, the only heat that contributes to the moisture evaporation in the product will be the dissipated fraction of its respiratory heat. Thus, considering the above, the transpiration rate in a fresh produce can be expressed as follows:

$$r_{H_2O} = \frac{q}{\lambda}\left(\frac{RT}{PM_{H_2O}}\right) + k(a_{wp} - a_{wat}) \qquad (6)$$

where r_{H2O} is the transpiration rate (in $cm^3\ kg^{-1}\ d^{-1}$), q is the total heat transferred to the product (kJ d^{-1}), λ is the latent heat of moisture evaporation (kJ kg^{-1}), R is the gas constant, T is the temperature, P is the pressure, and M_{H2O} is the water molar mass. In addition, k is total mass transfer coefficient due the water vapor differential ($cm^3\ kg^{-1}\ d^{-1}$), a_{wp} is the water activity in the product, and a_{wat} is the water activity in the surrounding atmosphere. The coefficient k will be influenced by temperature, and this relation can be expressed using the Arrhenius' Law [41]. If the packing system is not in thermal equilibrium, it may be assumed that the effective heat contributing to the evaporation of water besides the water activity gradient is the dissipated fraction of respiratory heat to simplify the calculations in Eq. (6). In that case, it is necessary to write the energy balance of the packaging system to estimate the temperature change in the product [61].

In case that the vapor pressure in the atmosphere (in the packaging headspace) becomes equal to the saturation pressure at the storage temperature, the water lost by the product as vapor will be condensed almost instantly inside the package [63]. On the other hand, as a result of transpiration, the weight of the packaged product will be decreasing over the storage time which can bring undesirable effects on its quality. This weight loss can be quite significant when the differences in water activities between the product and the surrounding atmosphere are large or when the storage time is long, and for this reason, it should be considered in the modeling of the MAP system [41].

2.1.3. Gas permeation through the packaging film

The permeability of a material to different gases will be determined by its molecular structure, thickness, surface area available for mass transfer, concentration gradient, pressure differential and temperature [13, 39, 64]. For some biopolymers, the relative humidity should also be considered [40]. When the film structure is uniform, without cracks or perforations, the gas will flow through the material by molecular diffusion. First, the gas will be adsorbed to the near face of the film, then the gas will diffuse through the film due to the concentration differential (from the point of greatest concentration to the point of lowest concentration), and thereafter, it will be desorbed from the farthest face from the film [12, 39].

The permeation or transfer of a gas i through the material constituting the packaging system can be expressed as follows [13, 65]:

$$J_{fi} = -\frac{Q_i A (p_{i-}p_{i,\,out})}{L} \frac{P}{RT} \tag{7}$$

where J_{fi} is the gas permeation through the continuous packaging film (mol d^{-1}), Q_i is the permeability coefficient of gas i in the packing material (cm^3 mm m^{-2} atm^{-1} d^{-1}), and p_i and $p_{i,out}$ are the partial pressures of gas i in and out of the package (atm). P is the local pressure (atm), A is the film surface area (m^2), and L is the thickness of the packaging material (mm). The permeability of the packaging films will depend on the storage temperature, and this can be represented by using the Arrhenius' Law considering the permeability coefficient Qi as a function of temperature [43].

2.1.4. Gas transmission through perforations in the packaging system

The gas transfer through perforations made in each packaging film can be described by using the Fick's diffusion equation with a correction term (ε in Eq. (8)) due to the resistance to diffusion at the perforation [13, 66]. Thus, the transmission rate of gas i through the perforation (K_{Tri}, in cm^3 d^{-1}) can be defined as follows:

$$K_{TRi} = \frac{D_i A_h}{L + \varepsilon} \tag{8}$$

where D_i is the diffusion coefficient of the packaging film to gas i, Ah is the cross-sectional area of the perforations, L is the film thickness, and ε is the correction term due the diffusion resistance in the perforation channel. The correction term ε can be estimated as approximately 0.5 times the effective diameter (d_e) of the perforations, the effective diameter being related to the diameter of each perforation and to the total number of perforations through the expression $d_e = N^{1/2}d$ [10, 13, 59].

The transmission rate of each gas through the perforation depends on the cross-sectional area and length (film thickness) of the perforation itself, the diffusivity of the gas in the air (gas mixture in the area to which the gas diffuses) and the concentration difference between the inside headspace and the external atmosphere [13, 48]. The dependence of the diffusion coefficients with respect to temperature can be described in the same way as the permeability coefficients using Arrhenius' law.

2.2. Balance equations for the MAP system

In the headspace of a MAP system, the change in the O_2, CO_2, H_2O and N_2 concentrations can be described by using the mass balance equations for each gas considering the respiration and transpiration rates, the gas transfer through the packaging material, the gas transmission through the perforations in the package and the changes in volume or pressure. To write the equations, it is necessary to consider whether the packaging system is flexible or rigid [13, 48, 50].

If the packaging system is flexible (such as a bag), the packaging headspace will be varying with the change in the number of moles of gases in the system, while the pressure can be considered constant and equal to the atmospheric pressure. In this case, the balance equations will be as follows:

For O_2,

$$V\frac{dy_{O_2}}{dt} = K_{TRO_2}(y_{O_{2Out}} - y_{O_2}) + \frac{APQ_{O_2}}{L}(y_{O_{2Out}} - y_{O_2}) - r_{O_2}W - y_{O_2}\frac{dV}{dt} \qquad (9)$$

For CO_2,

$$V\frac{dy_{CO_2}}{dt} = K_{TRCO_2}(y_{CO_{2Out}} - y_{CO_2}) + \frac{APQ_{CO_2}}{L}(y_{CO_{2Out}} - y_{CO_2}) + r_{CO_2}W - y_{CO_2}\frac{dV}{dt} \qquad (10)$$

For H_2O

$$V\frac{dy_{H_2O}}{dt} = K_{TRH_2O}(y_{H_2O_{out}} - y_{H_2O}) + \frac{APQ_{H_2O}}{L}(y_{H_2O_{out}} - y_{H_2O}) + r_{H_2O}W - y_{H_2O}\frac{dV}{dt} \qquad (11)$$

For N_2,

$$V\frac{dy_{N_2}}{dt} = K_{TRN_2}(y_{N_{2Out}} - y_{N_2}) + \frac{APQ_{N_2}}{L}(y_{N_{2Out}} - y_{N_2}) - y_{N_2}\frac{dV}{dt} \qquad (12)$$

where y_{O_2}, y_{CO_2}, y_{H_2O} and y_{N_2} are the O_2, CO_2, water vapor and N_2 concentrations in the packaging headspace, respectively; y_{O_2out}, y_{CO_2out}, y_{H_2Oout} and y_{N_2out} are the gas external concentrations, V is the headspace volume (cm^3), W is the product weight (kg), and t is the storage time (d).

The headspace volume change over time can be written as follows:

$$\frac{dV}{dt} = W(r_{CO_2} - r_{O_2} + r_{H_2O}) + \sum_{i=1}^{4}\left((y_{iOut} - y_i)\left(K_{TRi} + \frac{APQ_i}{L}\right)\right) \tag{13}$$

where i is O_2, CO_2, water vapor or N_2.

If the headspace atmosphere is saturated (RH = 100 % and y_{H2O} = P_s/P, where Ps is the saturation pressure), the change in the water vapor concentration with the storage time in the headspace will be zero, and the water evaporated in the product due to its transpiration (r_{H2O}) will be condensing inside the package. In this case, the Eq. (11) can be written as:

$$\frac{dy_{H_2O}}{dt} = 0 \tag{14}$$

In this case, the contribution of r_{H2O} to the volume change in the packaging headspace is not considered in Eq. (13).

If the packaging system is rigid, the packaging pressure will be varying with the change in the moles of gases in the system while the headspace volume can be considered constant along the storage time. In this case, the balance equations will be as follows:

For O_2,

$$V\frac{dy_{O_2}}{dt} = K_{TRO_2}(y_{O_{2Out}} - y_{O_2}) + \frac{APQ_{O_2}}{L}(y_{O_{2Out}} - y_{O_2})$$
$$+ \frac{N\pi d^4(P_{out} - P)y_{O_2}}{128\mu L_h} - r_{O_2}W - y_{O_2}\frac{V}{P}\frac{dP}{dt} \tag{15}$$

For CO_2,

$$V\frac{dy_{CO_2}}{dt} = K_{TRCO_2}(y_{CO_{2Out}} - y_{CO_2}) + \frac{APQ_{CO_2}}{L}(y_{CO_{2Out}} - y_{CO_2})$$
$$+ \frac{N\pi d^4(P_{out} - P)y_{CO_2}}{128\mu L_h} + r_{CO_2}W - y_{CO_2}\frac{V}{P}\frac{dP}{dt} \tag{16}$$

For H_2O

$$V\frac{dy_{H_2O}}{dt} = K_{TRH_2O}(y_{H_2O_{out}} - y_{H_2O}) + \frac{APQ_{H_2O}}{L}(y_{H_2O_{out}} - y_{H_2O})$$
$$+ \frac{N\pi d^4(P_{out} - P)y_{H_2O}}{128\mu L_h} + r_{H_2O}W - y_{H_2O}\frac{V}{P}\frac{dP}{dt} \tag{17}$$

For N_2,

$$V\frac{dy_{N_2}}{dt} = K_{TRN_2}(y_{N_{2Out}} - y_{N_2}) + \frac{APQ_{N_2}}{L}(y_{N_{2Out}} - y_{N_2})$$
$$+ \frac{N\pi d^4(P_{out} - P)y_{N_2}}{128\mu L_h} - y_{N_2}\frac{V}{P}\frac{dP}{dt} \tag{18}$$

where $P_{out}-P$ represents the difference in pressure between the packaging headspace and the external atmosphere, which results in a gas flow through the packaging system that can be described by using the Poiseuille's Law [50, 59]. N is the number of perforations, d is the perforation diameter, μ is the viscosity of the gas mixture (which can be considered as air), and L_h is the effective length of the perforations ($L + \varepsilon$).

The headspace pressure change over time can be written as follows:

$$\frac{dP}{dt} = P\left(W(r_{CO_2} - r_{O_2} + r_{H_2O}) + \sum_{i=1}^{4}\left((y_{iOut} - y_i)\left(K_{TRi} + \frac{APQ_i}{L}\right) + y_i \frac{N\pi d^4(P_{out} - P)}{128\mu L_h}\right)\right)$$

(19)

where i is again O_2, CO_2, water vapor or N_2.

As before, if the headspace atmosphere is saturated (RH = 100%), the change in the water vapor concentration with the storage time in the headspace will be zero, and the water evaporated in the product due to its transpiration will be condensing inside the package. In this case, the Eq. (17) will be equal to Eq. (14). In this case again, the contribution of r_{H2O} to the pressure change in the packaging headspace is not considered in Eq. (19).

The weight change in the packed product over time can be determined as follows [41]:

$$\frac{dW}{dt} = -Wr_{H_2O}\left(\frac{PM_{H_2O}}{RT}\right)$$

(20)

where W is the product weight, r_{H2O} is the transpiration rate, P is the packaging pressure, R is the gas constant, T is the temperature, and M_{H2O} is the water molar mass.

2.3. Model considerations and numerical solution

By using the mass balance equations for the MAP system in combination with the equations to represent the product respiration and transpiration and the transfer of the gases through the package and perforations, it is possible to represent the change in the composition of the packaging headspace along the time under defined conditions. However, for the set of equations to be useful and achieve a numerical solution, it is required to establish several considerations and assumptions in the MAP model [14, 41, 48, 50]:

- During the packaging time, it can be assumed that the respiration and transpiration rates in the product are not affected by ripening or senescence considering that in general, at this time, the product is expected to remain in a kind of pseudoequilibrium state while consuming the substrates that allow the development of these processes.

- After reaching the thermal equilibrium in the packaging system, this is conserved throughout the storage.

- All temperature dependent processes can be modeled based on Arrhenius' law.

- The stratification of gases inside the package is negligible considering that at the sizes of packaging usually used in MAP, the concentration gradients are small due to the similarity in diffusivities of the gases involved and the small lengths of gas diffusion.

- The flexible packaging system acts as a kind of cylinder-piston when there is a change in the composition and the amount of gas in the packaging headspace, keeping the pressure approximately constant. Also in the rigid packaging system, the volume remains approximately constant throughout the storage time.

- The amount of moisture gained by the product is negligible with respect to the amount lost due to its transpiration.

The equations for the MAP system can be solved numerically by using a suitable program routine to integrate multiple differential equations [13, 43, 48]. Whereas some components of the model solution, such as O_2 and CO_2 concentrations in headspace will change more rapidly than others such as water vapor concentration, headspace volume or product weight, the equations of the MAP system are characterized by being stiff or numerically unstable unless very small step sizes are used in the iteration process to find the solution. For this reason, it is recommended to use multistep implicit methods based on numerical differentiation formulas (NDF) [41]. When the packaging conditions such as concentration of gases and relative humidity in the external atmosphere, atmospheric pressure, product weight, packaging material, area and thickness of the flexible or rigid package and initial gas levels, headspace volume and pressure are defined, it is possible to estimate the change in the headspace atmosphere composition for a determined storage time.

3. Configuration of a suitable MAP

Each MAP design must be optimized for a specific product, since agricultural products have different metabolisms to each other, and as mentioned above, the MAP system must balance the processes of respiration, transpiration, ethylene production and gas permeation through the packaging that will be occurring simultaneously. Taking all of this into account, factors that affect both the metabolism of the packaged product and the gas permeation through the packaging should be considered for an optimal design of the MAP system [9, 14, 18]. Among these factors are firstly, the product maturity and ripeness stage, its weight and geometry and if it is whole or cut into fractions and secondly, the characteristics of the packaging system such as material, surface area, film thickness and headspace volume, and finally, the storage conditions such as temperature and external relative humidity [10, 13, 42]. With regard to the packaged product, it is also necessary to know in advance its transpiration, respiration rates at the packaging conditions and the optimum concentrations of these gases and relative humidity that increase its shelf life [10, 18, 43].

Once the rates of the metabolic processes and characteristics of the packaged product are known, it is also necessary to determine the gas levels and temperature that are most favorable for its preservation. In this way, it is possible to determine the configuration of the packaging system in which the different processes are balanced, and the appropriate gas levels are reached steadily over the storage time [10, 18, 67].

The design methodology of a successful MAP system can follow the steps shown in **Figure 1**.

First, it is necessary to define the product to be preserved and to determine its relevant characteristics such as maturity and ripeness stage, and configuration (whole or cut). It is then necessary to determine a suitable storage temperature that does not result in product deterioration (generally, it will be the most favorable temperature, but according to the specific conditions of distribution can be higher or even room temperature). Likewise, it is required to evaluate the levels of gases favorable for longer preserve the produce: O_2, CO_2 and HR. In order to determine these suitable conditions of temperature and gas concentrations, it may be required to perform experimental tests at various storage conditions or consult available data from literature for the product to be preserved.

To determine the gas transfer capacity required in the MAP system, it is necessary to pre-estimate the respiration and transpiration rates of the product to be packed. These rates can be determined as a function of gas levels and temperature by using Eqs. (4)–(6) and knowing previously the parameters of each equation that must be obtained experimentally or from other studies [41, 45, 47].

Once the aspects corresponding to the product to be preserved are known, then the configuration of the packaging system for its preservation is followed. First, it is necessary to evaluate of the packaging material to be used and estimation of its permeation rates to the different gases. This evaluation will of course depend on the transfer capacity required to properly balance the metabolism of the product. For a product with high respiration rates, a packing material with high permeability to O_2 and CO_2 is required. Likewise, for a product with high transpiration rate, a packing material with a high permeability to water vapor will be necessary. Other considerations must be made such as the availability and cost of the packaging material. Many biodegradable polymers such as PLA or EVOH are suitable for MAP systems but their availability on the market is still limited, and their cost is still high compared to synthetic materials such as PP and LDPE.

To determine the required transfer capacity in the package to reach the suitable gas levels (O_2, CO_2 or RH), it is necessary to estimate the transfer area of the packaging material and/or the size of the perforations required by using the mass balance equations of the MAP system in steady state [13, 43]. If the gas transfer capacity in the package is enough to compensate the produce respiration and transpiration, after a certain time an atmosphere of equilibrium will be reached inside the MAP system and for a flexible packaging, for example, Eqs. (9)–(12) may be rewritten as follows:

$$\left(K_{TRO_2} + \frac{APQ_{O_2}}{L}\right)\left(y_{O_{2Out}} - y_{O_{2\,eq}}\right) - r_{O_2}W - y_{O_{2\,eq}}\frac{dV}{dt} = 0 \tag{21}$$

$$\left(K_{TRCO_2} + \frac{APQ_{CO_2}}{L}\right)\left(y_{CO_{2Out}} - y_{CO_{2\,eq}}\right) + r_{CO_2}W - y_{CO_{2\,eq}}\frac{dV}{dt} = 0 \tag{22}$$

$$\left(K_{\text{TRH}_2\text{O}} + \frac{APQ_{\text{H}_2\text{O}}}{L} \right)\left(y_{\text{H}_2\text{O}_{\text{out}}} - y_{\text{H}_2\text{O}_{\text{eq}}} \right) + r_{\text{H}_2\text{O}}W - y_{\text{H}_2\text{O}_{\text{eq}}}\frac{dV}{dt} = 0 \tag{23}$$

where $y_{\text{O}_2\text{ eq}}$, $y_{\text{CO}_2\text{ eq}}$ and $y_{\text{HO}_2\text{ eq}}$ are the equilibrium O_2, CO_2 and water vapor concentrations. The change in the free package volume over time dv/dt, is substituted from Eq. (13). The N_2 balance is not necessary since N_2 concentration (molar fraction) is a function of the O_2, CO_2 and water vapor concentrations considering that the sum of all concentrations must be equal to 1, and that the presence of other gases is negligible. In the case of rigid packages, Eqs. (14)–(17)

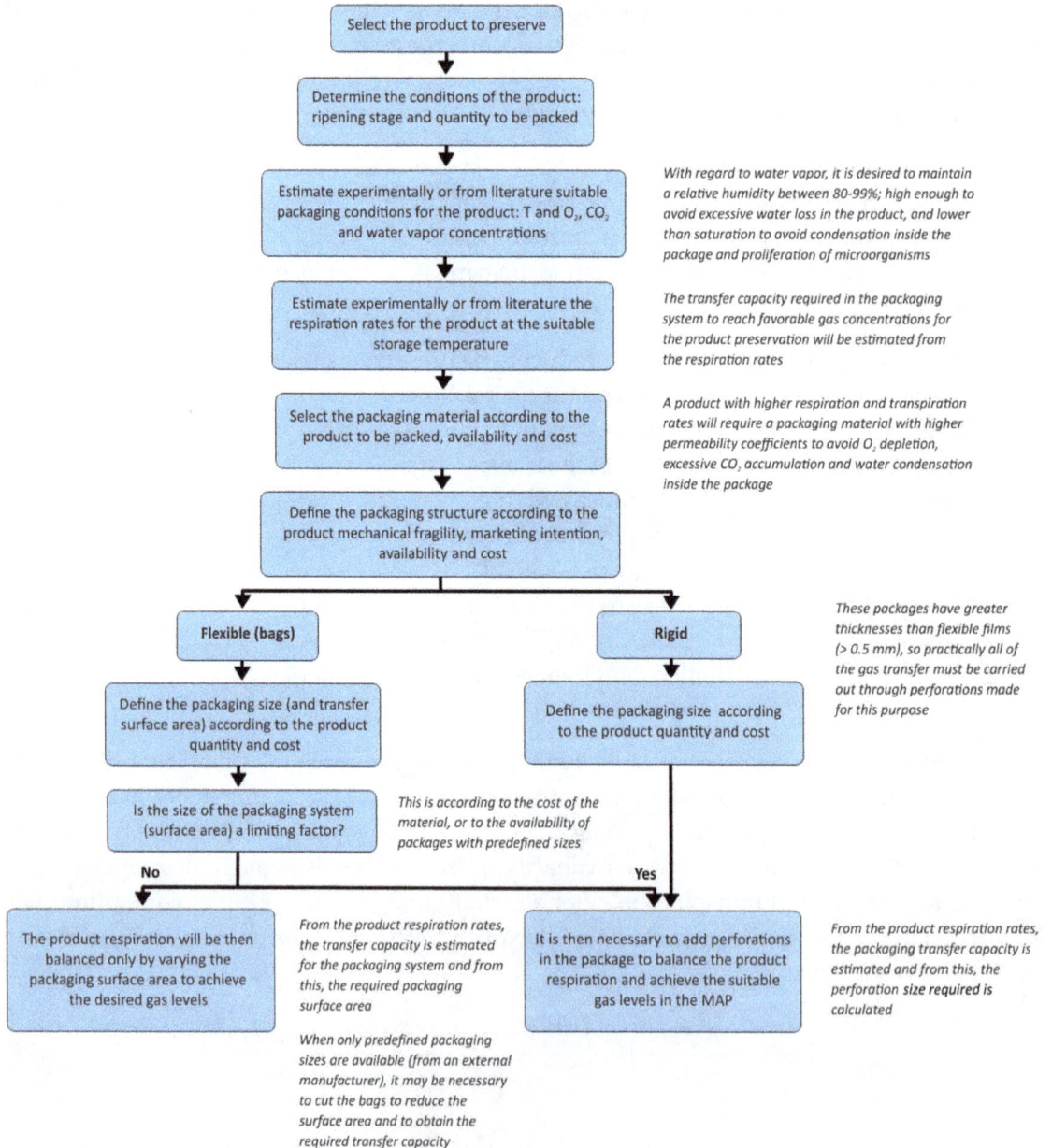

Figure 1. Schematic diagram for the configuration of a suitable MAP system.

can be rewritten similar to Eqs. (20)–(23) with the change in concentration of each gas with respect to time equal to zero in steady state. In that case, the change in the headspace pressure over time dP/dt is substituted from Eq. (19).

The meaning of Eqs. (21)–(23) is that to reach a determined steady concentration of gas, a given transfer capacity by the package will be required to match the produce respiration and transpiration. To achieve this transfer capacity, the type of packaging material, the package surface area and the number and size of perforations can be modified. By knowing the produce respiration and transpiration rates, it is possible define a packaging system to achieve a self-sustaining controlled atmosphere with favorable gas levels [13].

Depending whether the size of the package is defined or not, two possible configurations arise. In the first configuration, if the size is not defined, it is possible to achieve the required transfer capacity of the package simply by increasing the film surface area. In this case, the K_{TR} values of each gas are zero in the absence of perforations and four variables will be taken: the equilibrium O_2, CO_2 and water vapor concentrations, y_{O_2eq}, y_{CO_2eq}, y_{H_2Oeq} and the required surface area. As there are four variables and three independent equations, fixing one of the former, zero degrees of freedom are left, and the remaining three variables are determined in the calculation. What this means is that it is only possible to configure the system to specifically set only a gas concentration, that of O_2, CO_2 or water vapor. If it is wanted to set the equilibrium O_2 concentration in the MAP system, for example, the equilibrium CO_2 and water vapor concentrations (and of course the surface area) will be already defined and cannot be set. As is not possible to isolate A and $y_{O2\ eq}$ in Eqs. (21)–(23), it is necessary to calculate the value of the former (and the other concentrations) iteratively.

It is worth noting that the water vapor concentration in the packaging headspace will be about 1% (mole fraction) in almost all situations due to its low saturation pressure with respect to the atmospheric pressure at the typical storage conditions (e.g., at 10°C and 1 atm, 1.21% water vapor is equivalent to 100% RH). Considering this, it is possible to somehow simplify calculations to set a defined O_2 concentration and find the required area by assuming the water vapor concentration (between 1 and 2% mole fraction) and run iterations using only two equations instead of three.

In the second configuration, if the size of the package is defined, the surface area will be constant and to achieve the required transfer capacity is necessary to make perforations. In this case, the equilibrium gas concentrations, y_{O_2eq}, y_{CO_2eq}, y_{H_2Oeq} and the transmission coefficients through the perforation, K_{TRO_2}, K_{TRCO_2}, K_{TRH_2O} and K_{TRN_2}, will be taken as variables. However, from Eq. (8), it can be seen that just as the diffusion coefficients of each gas in the atmosphere, the thickness of the bag and the ε correction term are already defined, the real variables are the diameter and the number of perforations in the package. If the number of perforations is fixed, only the diameter is variable, and if the perforation diameter is fixed, only the number of these may be varied, although it is better to define the diameter to estimate an integer number of perforations [13]. Four variables will then be taken: the equilibrium O_2, CO_2 and water vapor concentrations, y_{O_2eq}, y_{CO_2eq}, y_{H_2Oeq} and the diameter or number of perforations. As in the first setting, when one of the concentrations is fixed, zero degrees of freedom are left, and the remaining concentration and the diameter or the number of perforations are determined in the iterative calculation.

In general, for n independent components in the packaging headspace, it will be $n + 1$ variables (counting the variable related to the transfer capacity in the package, e.g., the diameter of perforations), and n independent equations. By setting one of the gas concentrations, the system can be solved.

In the case of rigid packages where a higher transfer capacity is given by the perforations, the same procedure explained for the second configuration can be followed. Depending on the configuration defined for the MAP system and the initial gas concentrations in the headspace, it can take more or less time to reach a steady state and obtain the desired gas concentration. It can then be quite convenient to seal the package with an initial concentration close to the desired one to achieve equilibrium more quickly [13].

Author details

Diego A. Castellanos[1,2] and Aníbal O. Herrera[2*]

*Address all correspondence to: aoherreraa@unal.edu.co

1 Food Engineering Program, Agricultural University Foundation of Colombia, Bogotá, Colombia

2 Postharvest Laboratory, National University of Colombia, Bogotá, Colombia

References

[1] El-Ramady HR, Domokos-Szabolcsy É, Abdalla NA, Taha HS, Fári M. Postharvest management of fruits and vegetables storage. Sustainable Agriculture Reviews. 2015;**15**:65-152. DOI: 10.1007/978-3-319-09132-7_2

[2] Salveit M, editor. Commercial Storage of Fruits, Vegetables and Florist and Nursery Crops. Davis: University of California; 2005. p. 485

[3] Hörtensteiner S. Chlorophyll degradation during senescence. Annual Review of Plant Biology. 2006;**57**:55-77. DOI: 10.1146/annurev.arplant.57.032905.105212

[4] Lee SK, Kader AA. Preharvest and postharvest factors influencing vitamin C content of horticultural crops. Postharvest Biology and Technology. 2000;**20**(3):207-220. DOI: 10.1016/S0925-5214(00)00133-2

[5] Prasanna V, Prabha TN, Tharanathan RN. Fruit ripening phenomenon – An overview. Critical Reviews in Food Science and Nutrition. 2007;**47**:1-19. DOI: 10.1080/10408390600976841

[6] Herianus JD, Singh LZ, Tan S. Aroma volatiles production during fruit ripening of 'Kensington Pride' mango. Postharvest Biology and Technology. 2003;**27**:323-336. DOI: 10.1016/S0925-5214(02)00117-5

[7] Paliyath G, Murr D. Biochemistry of fruits. In: Paliyath G, Murr DP, Handa AK, Lurie S, editors. Postharvest Biology and Technology of Fruits, Vegetables and Flowers. Hoboken, New Jersey, USA: Wiley-Blackwell; 2008. pp. 19-50

[8] Wu C. An Overview of Postharvest Biology and Technology of Fruits and Vegetables. In 2010 AARDO Workshop on Technology on Reducing Post-harvest Losses and Maintaining Quality of Fruits and Vegetables; 2010. pp. 2-11. Available from: http://ir.tari.gov.tw:8080/bitstream/345210000/2813/1/public [Accessed: 2017-02-20]

[9] Caleb O, Mahajan P, Al-Said, Opara U. Modified Atmosphere Packaging Technology of Fresh and Fresh-cut Produce and the Microbial Consequences—A Review. Food and Bioprocess Technology. 2013;6:303-329. DOI: 10.1007/s11947-012-0932-4

[10] Del-Valle V, Hernández-Muñoz P, Catalá R, Gavara R. Optimization of an equilibrium modified atmosphere packaging (EMAP) for minimally processed mandarin segments. Journal of Food Engineering. 2009;91:474-481. DOI: 10.1016/j.jfoodeng.2008.09.027

[11] Sandhya KV. Modified atmosphere packaging of fresh produce: Current status and future needs. LWT - Food Science and Technology. 2010;43:381-392. DOI: 10.1016/j.lwt.2009.05.018

[12] Mangaraj S, Goswami T. Modified atmosphere packaging of fruits and vegetables for extending shelf-life: A review. Fresh Produce. 2009;3(1):1-31

[13] Castellanos DA, Cerisuelo JP, Hernández-Muñoz P, Herrera AO, Gavara R. Modelling the evolution of O_2 and CO_2 concentrations in MAP of a fresh product: Application to tomato. Journal of Food Engineering. 2016;168:84-95. DOI: 10.1016/j.jfoodeng.2015.07.019

[14] Mangaraj S, Goswami TK, Mahajan PV. Development and validation of a comprehensive model for map of fruits based on enzyme kinetics theory and Arrhenius relation. Journal of Food Science and Technology. 2015;52:4286-4295. DOI: 10.1007/s13197-014-1364-0

[15] Selcuk N, Erkan M. The effects of modified and palliflex controlled atmosphere storage on postharvest quality and composition of 'Istanbul' medlar fruit. Postharvest Biology and Technology. 2015;99:9-19. DOI: 10.1016/j.postharvbio.2014.07.004

[16] Farber J, Harris L, Parish M, Beuchat L, Suslow TV, Gorney JR, Garrett EH, Busta FF. Microbiological safety of controlled and modified atmosphere packaging of fresh and fresh-cut produce. Comprehensive Review in Food Science and Food Safety. 2003;2:142-160. DOI: 10.1111/j.1541-4337.2003.tb00032.x

[17] Bourke P, O'Beirne D. Effects of packaging type, gas atmosphere and storage temperature on survival and growth of Listeria spp. in shredded dry coleslaw and its components. International Journal of Food Science and Technology. 2004;39:509-523. DOI: 10.1111/j.1365-2621.2004.00811.x

[18] Castellanos DA, Polanía W, Herrera AO. Development of an equilibrium modified atmosphere packaging (EMAP) for feijoa fruits and modeling firmness and color evolution. Postharvest Biology and Technology. 2016;120:193-203. DOI: 10.1016/j.postharvbio.2016.06.012

[19] Soltani M, Alimardani R, Mobli H, Mohtasebi S. Modified atmosphere packaging: A progressive technology for shelf-life extension of fruits and vegetables. Journal of Applied Packaging Research. 2015;**7**:art. 2

[20] Lee D, Kang J, Renault P. Dynamics of internal atmospheres and humidity in perforated packages of peeled garlic cloves. International Journal of Food Science and Technology. 2000;**35**(5):455-464. DOI: 10.1046/j.1365-2621.2000.00397.x

[21] Artés F, Vallejo F, Martinez J. Quality of broccoli as influenced by film wrapping during shipment. European Food Research and Technology. 2001;**213**(6):480-483. DOI: 10.1007/s002170100390

[22] Artés F, Martinez J. Quality of cauliflower as influenced by film wrapping during shipment. European Food Research and Technology. 1999;**209**(5):330-334. DOI: 10.1007/s002170050504

[23] Serrano M, Martinez-Romero D, Guillen F, Castillo S, Valero D. Maintenance of broccoli quality and functional properties during cold storage as affected by modified atmosphere packaging. Postharvest Biology and Technology. 2006;**39**(1):61-68. DOI: 10.1016/j.postharvbio.2005.08.004

[24] Plestenjak A, Pozrl T, Hribar J, Unuk T, Vidrih R. Regulation of metabolic changes in shredded cabbage by modified atmosphere packaging. Food Technology and Biotechnology. 2008;**46**(4):427-433

[25] Karakas B, Yildiz F. Peroxidation of membrane lipids in minimally processed cucumbers packaged under modified atmospheres. Food Chemistry. 2007;**100**(3):1011-1018. DOI: 10.1016/j.foodchem.2005.10.055

[26] Liu F, Li Y. Storage characteristics and relationships between microbial growth parameters and shelf life of MAP sliced onions. Postharvest Biology and Technology. 2006;**40**(3):262-268. DOI: 10.1016/j.postharvbio.2006.01.012

[27] Allende A, Luo Y, McEvoy J, Artés F, Wang C. Microbial and quality changes in minimally processed baby spinach leaves stored under super atmospheric oxygen and modified atmosphere conditions. Postharvest Biology and Technology. 2006;**33**:51-59. DOI: 10.1016/j.postharvbio.2004.03.003

[28] Domínguez I, Lafuente MT, Hernández-Muñoz P, Gavara R. Influence of modified atmosphere and ethylene levels on quality attributes of fresh tomatoes (*Lycopersicon esculentum* Mill.). Food Chemistry. 2016;**209**:211-219. DOI: 10.1016/j.foodchem.2016.04.049

[29] Hertog MLATM, Nicholson SE, Banks NH. The effect of modified atmospheres on the rate of firmness change in 'Braeburn' apples. Postharvest Biology and Technology. 2001;**23**:175-184. DOI: 10.1016/S0925-5214(01)00126-0

[30] Santos CMS, de Barros Vilas Roas EV, Botrel N, Marques AC. Influência da atmósfera controlada sobre a vida pos-colheita e qualidad de banana 'Prata Ana'. Ciencia e Agrotecnologia. 2006;**30**(2):317-322. DOI: 10.1590/S1413-70542006000200018

[31] Villalobos M, Serradilla M, Martín A, Ruiz-Moyano S, Pereira C, Córdoba M. Use of equilibrium modified atmosphere packaging for preservation of 'San Antonio' and 'Banane' breba crops (*Ficus carica* L.). Postharvest Biology and Technology. 2014;**98**:14-22. DOI: 10.1016/j.postharvbio.2014.07.001

[32] Almenar E, Samsudin H, Auras R, Harte B, Rubino M. Postharvest shelf life extension of blueberries using a biodegradable package. Food Chemistry. 2008;**110**:120-127. DOI: 10.1016/j.foodchem.2008.01.066

[33] Candir E, Ozdemir A, Kamiloglu O, Soylu E, Dilbaz R, Ustun D. Modified atmosphere packaging and ethanol vapor to control decay of 'Red Globe' table grapes during storage. Postharvest Biology and Technology. 2012;**63**:98-106. DOI: 10.1016/j.postharvbio.2011.09.008

[34] Mangaraj S, Goswami T, Giri S, Joshy C. Design and development of a modified atmosphere packaging system for guava (cv. Baruipur). Journal of Food Science and Technology. 2014;**55**(11):2925-2946. DOI: 10.1007/s13197-012-0860-3

[35] Waghmare R, Annapure U. Combined effect of chemical treatment and/or modified atmosphere packaging (MAP) on quality of fresh-cut papaya. Postharvest Biology and Technology. 2013;**85**:147-153. DOI: 10.1016/j.postharvbio.2013.05.010

[36] Cheng Y, Liu L, Zhao G, Shen C, Yan H, Guan J, Yang K. The effects of modified atmosphere packaging on core browning and the expression patterns of PPO and PAL genes in 'Yali' pears during cold storage. LWT - Food Science and Technology. 2015;**60**:1243-1248. DOI: 10.1016/j.lwt.2014.09.005

[37] Banda K, Caleb O, Jacobs K, Opara U. Effect of active-modified atmosphere packaging on the respiration rate and quality of pomegranate arils (cv. Wonderful). Postharvest Biology and Technology. 2015;**109**:97-105. DOI: 10.1016/j.postharvbio.2015.06.002

[38] Zhang M, Xiao G, Peng J, Salokhe V. Effects of single and combined atmosphere packages on preservation of strawberries. International Journal of Food Engineering. 2005;**1**(4):1556-3758. DOI: 10.2202/1556-3758.1043

[39] Mangaraj S, Goswami T, Mahajan P. Applications of plastic films for modified atmosphere packaging of fruits and vegetables: A review. Food Engineering Reviews. 2009;**1**:133-158. DOI: 10.1007/s12393-009-9007-3

[40] Auras R, Almenar E. Permeation, sorption and diffusion in poly(lactic acid). In: R. Auras, L. Lim T, Selke SEM, Tsuji H, editors. Poly(Lactic Acid): Synthesis, structures, properties, processing and application. New York: Wiley & Sons; 2010. pp. 164-170. DOI: 10.1002/9780470649848.ch12

[41] Castellanos DA, Herrera DR, Herrera AO. Modelling water vapour transport, transpiration and weight loss in a perforated modified atmosphere packaging for feijoa fruits. Biosystems Engineering. 2016;**151**:190-200. DOI: 10.1016/j.biosystemseng.2016.08.015

[42] Del-Nobile ME, Licciardello F, Scrocco C, Muratore G, Zappa M. Design of plastic packages for minimally processed fruits. Journal of Food Engineering. 2007;**79**:217-224. DOI: 10.1016/j.jfoodeng.2006.01.062

[43] Mahajan PV, Oliveira FAR, Montanez JC, Frias J. Development of user-friendly software for design of modified atmosphere packaging for fresh and fresh-cut produce. Innovative Food Science and Emerging Technologies. 2007;8:84-92. DOI: 10.1016/j.ifset.2006.07.005

[44] Bovi G, Caleb O, Linke M, Rauh C, Mahajan P. Transpiration and moisture evolution in packaged fresh horticultural produce and the role of integrated mathematical models: A review. Biosystems Engineering. 2016;150:24-39. DOI: 10.1016/j.biosystemseng.2016.07.013

[45] Fonseca S, Oliveira F, Brecht J. Modelling respiration rate of fresh fruits and vegetables for modified atmosphere packages: A review. Journal of Food Engineering. 2002;52:99-119. DOI: 10.1016/S0260-8774(01)00106-6

[46] Lu L, Tang Y, Lu S. A kinetic model for predicting the relative humidity in modified atmosphere packaging and its application in Lentinula edodes packages. Mathematical Problems in Engineering. 2013;304016. DOI: 10.1155/2013/304016

[47] Mendoza R, Castellanos DA, García JC, Vargas JC, Herrera AO. Ethylene production, respiration and gas exchange modelling in modified atmosphere packaging for banana fruits. 2016;51(3):777-788. DOI: 10.1111/ijfs.13037

[48] González-Buesa J, Ferrer-Mairal A, Oria R, Salvador ML. A mathematical model for packaging with microperforated films of fresh-cut fruits and vegetables. Journal of Food Engineering. 2009;95:158-165. DOI: 10.1016/j.jfoodeng.2009.04.025

[49] Li L, Xi-Hong L, Zhao-Jun B. A mathematical model of the modified atmosphere packaging (MAP) system for the gas transmission rate of fruit produce. Food Technology and Biotechnology. 2010;48(1):71-78

[50] Xanthopoulos G, Koronaki E, Boudouvis A. Mass transport analysis in perforation-mediated modified atmosphere packaging of strawberries. Journal of Food Engineering. 2012;111:326-335. DOI: 10.1016/j.jfoodeng.2012.02.016

[51] Taiz L, Zeiger E, Møller IM, Murphy A. Plant Physiology and Development. 6th ed. Sunderland: Sinauer Associates, Inc.; 2015. p. 761

[52] Mir N, Beaudry R. Atmosphere control using oxygen and carbon dioxide. In: Knee M, editor. Fruit Quality and Its Biological Basis. Boca Raton: Sheffield Academic; 2002. pp. 122-156

[53] Mahajan P, Luca A, Edelenbos M. Development of a small and flexible sensor-based respirometer for real-time determination of respiration rate, respiratory quotient and low O_2 limit of fresh produce. Computer and Electronics in Agriculture. 2016;121:347-353. DOI: 10.1016/j.compag.2015.12.017

[54] Tano K, Kamenan A, Arul J. Respiration and transpiration characteristics of selected fresh fruits and vegetables. Agronomie Africaine. 2005;17(2):103-115

[55] Bhande SD, Ravindra MR, Goswami TK. Respiration rate of banana fruit under aerobic conditions at different storage temperatures. Journal of Food Engineering. 2008;87(1):116-123. DOI: 10.1016/j.jfoodeng.2007.11.019

[56] Finnegan E, Mahajan P, O'Connell M, Francis G, O'Beirne D. Modelling respiration in fresh-cut pineapple and prediction of gas permeability needs for optimal modified

atmosphere packaging. Postharvest Biology and Technology. 2013;**79**:47-53. DOI: 10.1016/j.postharvbio.2012.12.015

[57] Wang ZW, Duan HW, Hu CY. Modelling the respiration rate of guava (*Psidium guajava* L.) fruit using enzyme kinetics, chemical kinetics and artificial neural network. European Food Research and Technology. 2009;**229**:495-503. DOI: 10.1007/s00217-009-1079-z

[58] Heydari A, Shayesteh K, Eghbalifam N, Bordbar H, Falahatpisheh S. Studies on the respiration rate of banana fruit based on enzyme kinetics. International Journal of Agriculture & Biology. 2010;**12**(1):145-149

[59] Kwon MJ, Jo YH, An DS, Lee DS. Applicability of simplified simulation models for perforation-mediated modified atmosphere packaging of fresh produce. Mathematical Problems in Engineering. 2013;267629. DOI: 10.1155/2013/267629

[60] Mangaraj S, Goswami TK. Measurement and modeling of respiration rate of guava (CV. Baruipur) for modified atmosphere packaging. International Journal of Food Properties. 2011;**14**:609-628. DOI: 10.1080/10942910903312403

[61] Song Y, Vorsa N, Yam KL. Modeling respiration-transpiration in a modified atmosphere packaging system containing blueberry. Journal of Food Engineering. 2002;**53**:103-109. DOI: 10.1016/S0260-8774(01)00146-7

[62] Sousa-Gallagher M, Mahajan PV, Mezdad T. Engineering packaging design accounting for transpiration rate: Model development and validation with strawberries. Journal of Food Engineering. 2013;**119**:370-376. DOI: 10.1016/j.jfoodeng.2013.05.041

[63] Rennie T, Tavoularis S. Perforation-mediated modified atmosphere packaging Part I. Development of a mathematical model. Postharvest Biology and Technology. 2009;**51**(1):1-9. DOI: 10.1016/j.postharvbio.2008.06.007

[64] Techavises N, Hikida Y. Development of a mathematical model for simulating gas and water vapor exchanges in modified atmosphere packaging with macroscopic perforations. Journal of Food Engineering. 2008;**85**:94-104. DOI: 10.1016/j.jfoodeng.2007.07.014

[65] Geankoplis CJ. Transport Processes and Unit Operations. 3rd ed. New Jersey: Prentice Hall; 1993. pp. 408-413

[66] Chung D, Papadakis SE, Yam KL. Simple models for evaluating effects of smalls leaks on the gas barrier properties of food packages. Packaging Technology and Science. 2003;**16**:77-86. DOI: 10.1002/pts.616

[67] Mistriotis A, Briassoulis D, Giannoulis A, D'Aquino S. Design of biodegradable bio-based equilibrium modified atmosphere packaging (EMAP) for fresh fruits and vegetables by using micro-perforated poly-lactic acid (PLA). Posthrvest Biology and Technology. 2016;**111**:380-389. DOI: 10.1016/j.postharvbio.2015.09.022

5

Effects of Size, Storage Duration, and Modified Atmosphere Packaging on Some Pomological Characteristics of Wonderful Pomegranate Cultivar

Hatice Serdar and Serhat Usanmaz

Additional information is available at the end of the chapter

Abstract

Pomegranate fruit is very susceptible to storage conditions, and it losses weight and quality during long storage periods. The aim of the present study was to determine the effects of modified atmosphere packaging (MAP), storage duration, and fruit size on some quality attributes of Wonderful pomegranate cultivar. Results indicated that there is a clear relationship between the fruit weight and number of arils (R^2, 0.948) as expected. On the other hand, a significant relationship also exists between the number of arils and aril weight (R^2, 0.973). According to the results, it is also possible to estimate the juice content of pomegranate fruits by its aril weight (R^2, 0.994) and also from the fruit weight. Results also showed that modified atmosphere packaging has a clear effect on the protection of fruit weight. The loss in the fruit weight after 90 days of storage is found to be between 6.70 and 14.28% without MAP and between 2.64 and 6.24% with MAP. Similar results have been determined for aril weight and juice content. The other important result of the present study is that fruit size significantly affects the weight loss. The bigger fruits showed higher weight loss. No significant effect has been determined on the total soluble solids content of fruits for neither different sizes nor different storage conditions.

Keywords: pomegranate, Wonderful, storage duration, modified atmosphere packaging, fruit size

1. Introduction

Pomegranate plant is among the oldest known cultivated crops. According to Ref. [1], domestication of pomegranate tree dates back to 3000–4000 BC in the North of Iran and Turkey. Pomegranate fruits are classified under the group of berries. Diverse number of arils are found within the pomegranate fruit wrapped to the inside of leathery peel [2]. Pomegranates are reported to be originated from Central Asia [3, 4]. However, pomegranate tree is adaptable to a wide range of climate and soil conditions. The trees can grow in many different geographical places including the Mediterranean Basin, California, and Asia.

Pomegranate fruits are traditionally known to be very beneficial for human health, but its consumption was limited because of the hassle of extracting the arils from the fruit. According to Ref. [5], Hippocrates (400 BCE) used pomegranate extracts for many purposes, such as aid to digestion and eye inflammation. Since the beginning of the twenty-first century, many scientific studies have been done about the health benefits of pomegranate. Besides to fruits, flowers, bark, and leaves of pomegranates may contain beneficial phytochemicals which are antioxidant and antimicrobial, reduce blood pressure, and act against serious diseases such as cancer and diabetes [6].

Confirmation of the health benefits of pomegranates caused an improvement in the public awareness, and this caused an increase in the pomegranate consumption in the world. Therefore, production of this highly beneficial fruit started to increase to meet the increasing demand. Pomegranate plant is produced throughout the world in subtropical and tropical areas. The harvesting period of pomegranate extends from August to November in the northern hemisphere and from March to May in the southern hemisphere. There is a year-round demand for pomegranate fruit throughout the world and its global availability is being met by postharvest storage [2]. However, pomegranate fruit is very susceptible to storage conditions, and it losses weight and quality during long storage periods. The main challenge of pomegranate fruit producers and marketers is maintaining postharvest quality. One of the main postharvest problems of pomegranate fruit is weight loss. Long and inappropriate storage causes huge losses not only on weight but also on fruit quality and income of the farmers. Weight loss also causes hardening of the husk and browning of the rind in which they reduce the attractiveness of fruits [7]. The weight loss of pomegranate fruit is reported to reach up to 32% in 8 weeks if stored at 22°C [8]. Many scientific studies were conducted about the favorable conditions for pomegranate storage in terms of prevention of weight loss, and they suggest 5–7°C and >90% relative humidity for the reduction of weight loss [9–11].

Modified atmosphere packaging (MAP) is a way of extending the shelf life of fresh food products. It helps to protect postharvest quality while reducing the weight loss. In a MAP system, there is a change in the concentration of gases in the packaging headspace due to the dynamic interaction between the metabolic and biochemical processes of the packaged product. When fresh produce respire, O_2 is consumed, and CO_2, ethylene, and water vapor are generated; and transfer of all of these gases through the packaging is regulated in MAP. The aim of the system is to balance these two processes to provide favorable conditions to preserve the product [7, 12, 13]. MAP is reported to be successful for the prolongation of storage duration of pomegranate fruits up to 3 months while reducing weight loss [8, 14, 15]. However, according to authors' knowledge, no studies were conducted about the combined effects of fruit size and modified atmosphere packaging. In the light of this information, the present study was aimed to determine the combined effects of fruit size, storage duration, and storage conditions on the pomological characteristics of Wonderful pomegranate cultivar.

2. Materials and methods

2.1. Materials

The main material of the present study was the Wonderful cultivar pomegranate fruits. This cultivar originated in Florida. The fruits are large when compared with most of other varieties/cultivars with a color of deep purple-red fruits. The fruit's inside is a deep crimson in color. The fruit is juicy and has a tart taste. The seeds are not very hard. This cultivar is better for fresh eating and juicing. It is a leading commercial cultivar in California and is the most popular cultivar in the world. It has a ripening Brix of 17–21%. The fruit requires about 160–180 days from flowering to ripening [2]. Fruits of Wonderful cultivar pomegranate were harvested at commercial maturity in October 2016 from 8 years of pomegranate trees grown in Cyprus. After harvest, fruits were classified by professional workers according to EU standards, and "extra" quality fruits were used in the experiments.

The other material of the study was the modified atmosphere packaging (MAP) bags. The MAP materials are purchased from Dekatrend Ltd. with the brand of Trendlife®. The material allows the CO_2 composition inside the package to be 9–15% and O_2 composition 6–10%. Trendlife® bags are reported to be manufactured from a semipermeable film which controls gas exchange. The film is based on the activity of several intelligent molecules placed inside the film. These molecules allow O_2 to enter the package at a rate offset by the consumption O_2 by the commodity. Similarly, the film makes it possible for CO_2 to be vented from the package to offset the production of CO_2 by the commodity. The film, on the other hand, vents the released ethylene gas from the package, thereby eliminating the possibility of acceleration of repining, senescence, deterioration, and decay. It also enriches the modified atmosphere with sufficient relative humidity and completely inhibits the weight loss.

2.2. Experimental design, treatments, and measurements

Fruits with five different sizes (4″, 6″, 8″, 10″, and 12″) were subjected to three different storage durations (*the main plots*: 0, 45, and 90 days) and two different storage conditions (*the subplots*: control and modified atmosphere packaging). The fruit sizes are standard sizes for boxes with dimensions of 30 × 40 × 12 cm. Experiments were carried out with five different sizes of fruits with three replicates for each, and fruits were arranged in split-split plot design in a cold store with 5–7°C and 90–95% relative humidity. At the first day of harvest, fruit weight (g), fruit diameter (mm), number of arils, aril weights (g/100 arils), juice content (ml/100 arils), and total soluble solids (TSS) content (%) were measured and noted for the 0 day main factor. For the other two factors (45 and 90 days), fruit diameters (mm) and fruit weights (g) were measured at the first day of harvest and noted. After 45 and 90 days of storage, either with or without MAP, other measurements (fruit weight, fruit diameter, aril weights, juice content, and TSS) were done.

2.3. Statistical analysis

Data from the experiments were subjected to regression analyses to describe the relationship between fruit weight, number of arils, aril weight, and juice content. To determine the effects of storage condition and storage duration on the quality attributes of pomegranate fruits, the data of the experiments were subjected to an ANOVA, and mean separations were done using Duncan's multiple range test at $P < 0.05$.

3. Results and discussions

3.1. Relationship between pomological characteristics of pomegranate fruits at harvest

A significant difference has been determined among the fruit sizes for five out of six tested pomological characteristics of pomegranate fruits at harvest (**Table 1**). The results for fruit weight and fruit width are expected due to the nature of the fruit size. As the fruit weight and width increase, the size decreases. Size represents the number of fruits present in the same box. The smallest fruit width and weight are noted for size 12 with 80.2 ± 1.00 mm and 250 ± 20.7 g, respectively. The highest fruit width and weight were determined for size 4 with 118.1 ± 3.09 mm and 929 ± 50.3 g, respectively. Ref. [16] reported that the fruit weight of 12 cultivars ranged from 103.4 to 505.0 g in Iran. In another study, Ref. [17] noted that the average fruit weight of four different varieties ranged from 241.1 to 319.8 g in Turkey. Fruit size is much related with cultivars/varieties, environmental conditions, and growing conditions. Therefore, it is not surprising to have diverse fruit weights.

On the other hand, a clear difference has been determined for the number of arils per fruit among the different sizes. It is an important result for farmers which shows that the fruit size is mainly affected by the number of arils present in the fruits. This proves that the adequate fertilization and pollination during flowering and fruit set are very important for fruits to include an adequate number of cells, which turns into arils during development. According to Ref. [18], some fruits have fewer than 300 arils per fruit and some have up to 985 in the largest fruits. Ref. [4] reported that the production of large fruits requires adequate numbers of both functional ovules and a source of viable pollen on flowers. Ovule differentiation in pomegranates occurs before the opening of flower buds. For this reason, adequate pollination is crucial for fruit size and development.

Sizes	Fruit width (mm)	Fruit weight (g)	Number of arils	Aril weight (g/100 arils)	Juice content (ml/100 arils)	TSS (% Brix)
Size 12	80.2 ± 1.00 (e)	250 ± 20.7 (e)	331 ± 31.2 (e)	35.7 ± 3.06 (d)	19.3 ± 2.08 (d)	17.8 ± 0.76 (a)
Size 10	86.4 ± 1.89 (d)	354 ± 24.5 (d)	476 ± 32.7 (d)	43.3 ± 1.53 (c)	26.7 ± 1.53 (c)	17.9 ± 0.85 (a)
Size 8	95.6 ± 3.75 (c)	449 ± 18.4 (c)	589 ± 25.7 (c)	47.7 ± 3.06 (bc)	30.5 ± 2.18 (b)	17.5 ± 0.50 (a)
Size 6	112.0 ± 2.47 (b)	680 ± 68.6 (b)	752 ± 30.0 (b)	51.7 ± 2.31 (ab)	33.0 ± 2.69 (b)	17.7 ± 0.68 (a)
Size 4	118.1 ± 3.09 (a)	929 ± 50.3 (a)	860 ± 18.1 (a)	55.0 ± 2.00 (a)	37.1 ± 0.85 (a)	17.5 ± 0.50 (a)

Values followed by the same letter or letters are not significantly different at a 5% level (Duncan's multiple range test).

Table 1. Summary of some pomological data of pomegranate fruits at harvest.

Moreover, a significant difference has been determined among the aril weights of fruits for different sizes. This, on the other hand, shows that not only the number of arils but also the weight of arils is affecting the fruit size. No significant difference has been determined for the total soluble solids content of the fruits with different sizes.

At the first day of harvest, a clear relationship has been determined between the fruit weight and the aril weight (**Figure 1**). The adjusted R^2 is calculated as 0.866. The results showed that as the fruit weight increases, the aril weight also increases. It is now possible to estimate the aril weight of a Wonderful cultivar pomegranate fruit at harvest by using the equation $y = 0.025x + 32.95$. In another study by Ref. [18], the relationship between fruit weight and total aril weight is described with the equation of $y = 0.525x - 6.725$ ($R^2 = 0.951$). Similar relationship has been determined between the fruit weight and the juice content. The adjusted R^2 is calculated as 0.858 for this relationship.

On the other hand, an important linear relationship has been determined between the fruit weight and the number of arils. One can estimate the number of arils of a Wonderful cultivar pomegranate fruit from its weight by using the equation $y = 0.754x + 199.2$ ($R^2 = 0.948$). Ref. [18] reported less relationship between the fruit weight and the number of arils with the equation of $y = 1.258x + 53.715$ ($R^2 = 0.744$). The variance of the fruit weight of the present study is higher than the variance of the study carried out by Ref. [18] which could be the main reason of the difference among the equations. Higher variation for the fruit weight increased

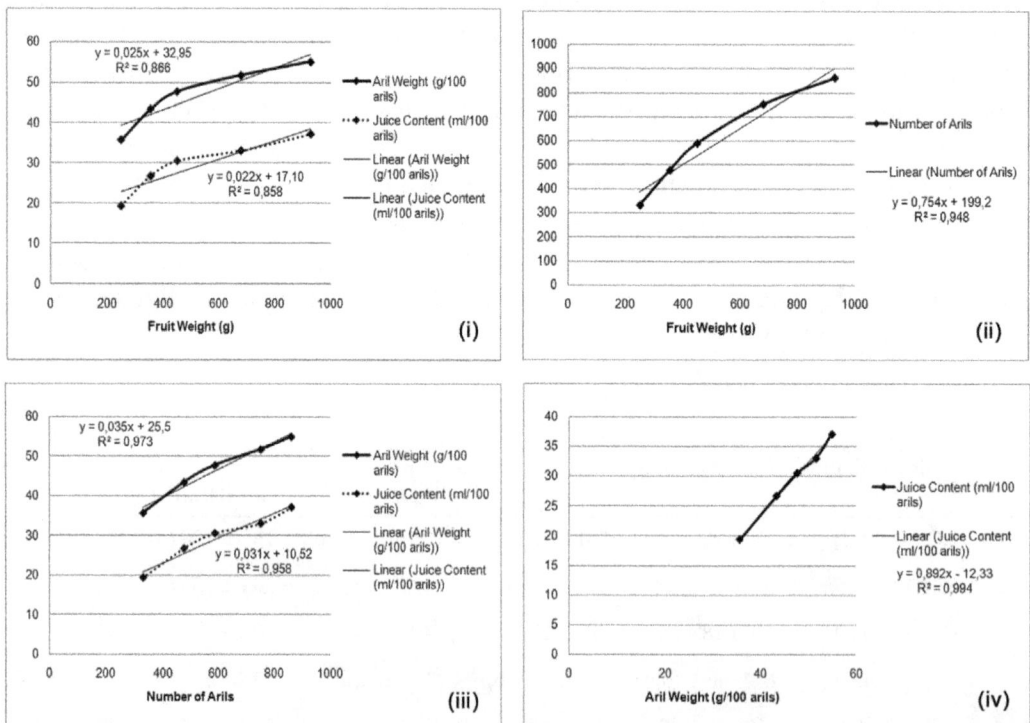

Figure 1. Relationship between fruit characteristics in Wonderful cultivar pomegranate: (i) fruit weight vs. aril weight and juice content, (ii) fruit weight vs. number of arils, (ii) number of arils vs. aril weight or juice content, and (iv) aril weight vs. juice content.

the reliability of the present study by having an improved adjusted R^2. Since there is a clear relationship between the fruit weight and the number of arils, a significant linear relationship has been determined for the number of arils vs. the aril weight or juice content.

3.2. Relationship between pomological characteristics of pomegranate fruits after 90 days of storage

Storage of fruits with or without MAP for 90 days caused a nonsignificant change in the relationship between fruit weight and aril weight (**Figure 2**). At the first day of harvest, the adjusted R^2 for a linear regression was 0.866 which decreased to 0.854 with MAP and to 0.825 without MAP. The relationship is found to be higher at the fruits stored with MAP than the fruits stored without MAP. Similarly, the relationship between the aril weight and the juice content had no significant change in 90 days of storage with or without MAP. These results confirmed that the changes in aril weight and juice content are in accordance with the changes in fruit weight. Although a significant change has been determined in fruit weight, aril weight, and juice content (**Figure 3**), those changes were found to be in accordance with each other.

3.3. Effects of storage conditions on some pomological characteristics of pomegranate fruit

Pomegranate fruit is very susceptible to weight loss due to the high porosity on the fruit, and its susceptibility is depending on storage conditions [9]. The present study revealed the effects of modified atmosphere packaging on the prevention of weight loss of pomegranate fruits. After 90 days of storage, a significant difference has been determined on the percent reduction of weight loss between fruits stored without MAP and fruits stored with MAP (**Figure 3**). It is clear from the figure that the modified atmosphere packaging protects the fruit weight. The loss in the fruit weight without MAP is found to be between 6.70 and 14.28%. The results showed that the fruit size is also affecting weight loss. The bigger fruits showed higher losses. The weight loss of the biggest fruits (size 4) is found to be more than twofold of the smallest size (size 12), for both with MAP and without MAP storage. On the other hand, MAP significantly protected the weight of the fruits. The weight loss of fruits with MAP is found to be between only 2.64 and 6.24% in 90 days of storage. Findings of the present study are in agreement with Ref. [19] where they

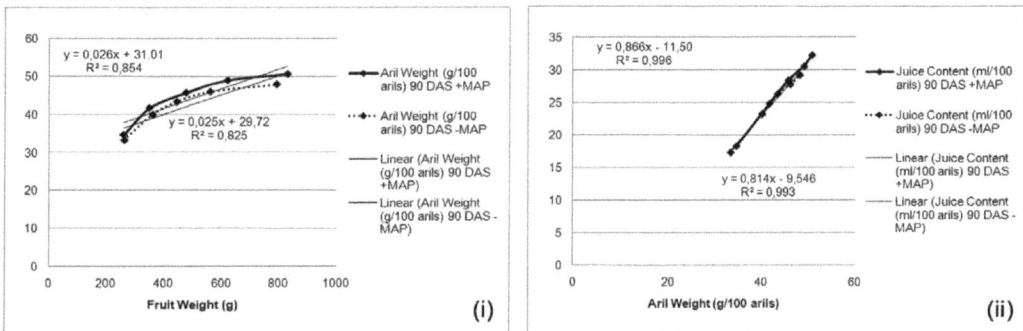

Figure 2. Relationship between fruit characteristics in Wonderful cultivar pomegranate 90 days after storage: (i) fruit weight vs. aril weight and (ii) aril weight vs. juice content.

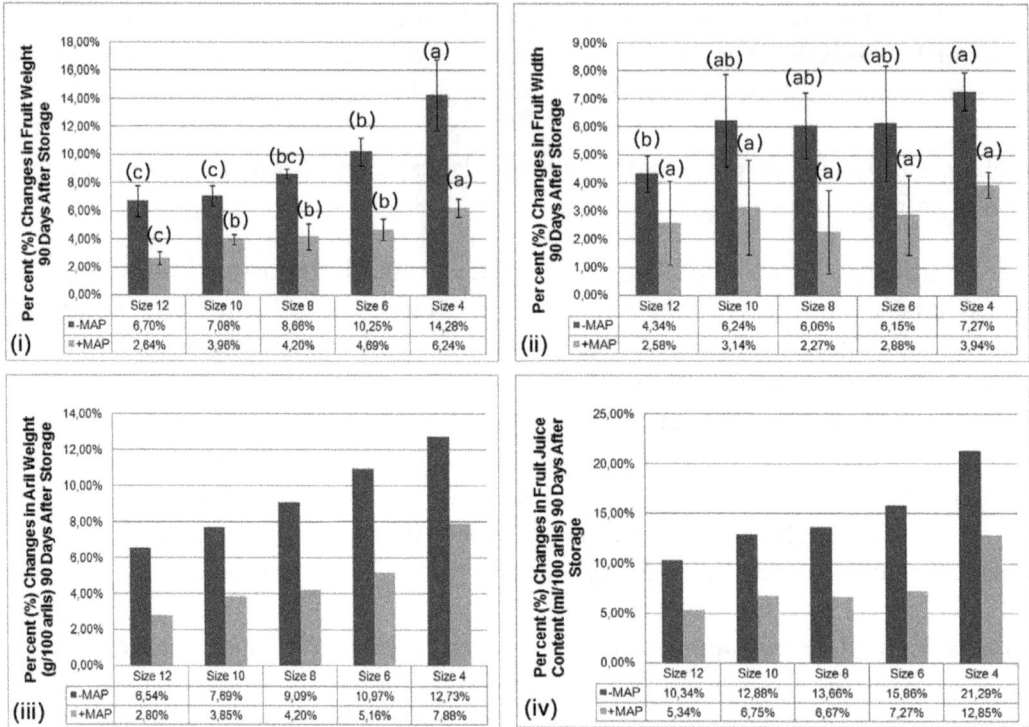

Figure 3. Percent (%) changes in fruit characteristics in Wonderful cultivar pomegranate 90 days after storage: (i) fruit weight, (ii) fruit width, (iii) aril weight, and (iv) juice content.

reported that the physical parameters of pomegranates were better when fruits were wrapped and refrigerated after 20 days of storage. They reported a weight loss for EPE-foam and polyethylene-film wrapped fruits as 0.8 and 0.98% which is about 4.97% for non-wrapped fruits at 20 days of storage at 7.5°C. On the other hand, Refs. [20, 21] also noted that modified atmosphere packaging has a clear influence on the protection of the weight of pomegranate fruits.

No significant difference has been found for the fruit widths of pomegranate fruits after 90 days of storage among different sizes. But, on the other hand, there is a significant influence of modified atmosphere packaging packaging on the fruit width, which protected the fruit width. The percent reduction in the fruit width is found to be between 4.34 and 7.27% for control fruits and between 2.58 and 3.94% for the fruits stored with MAP. The percent reduction in aril weight is found to be between 6.54 and 17.73% for the fruits stored without MAP for 90 days. On the other hand, the fruits with same sizes which stored under MAP conditions had only 2.80–7.88% reduction in the aril weight. These results showed that the MAP has a clear effect on the protection of aril weight as in fruit weight. The other important result of the present work is that the percent reduction in aril weight is increasing as the fruit size decreases (or fruit weight increases). The percent (%) reduction in juice content is found to be between 10.34 and 21.29% for the fruits stored without MAP for 90 days. On the other hand, the fruits with same sizes which stored under MAP conditions had only 5.34–12.85% reduction in the juice content. These results showed that the MAP has a clear effect on the protection of juice content. These results are in accordance with the results for aril weight as expected.

4. Conclusions

The results of the present study showed that there is a linear relationship between the fruit weight and the number of arils (R^2, 0.948), the number of arils and the aril weight (R^2, 0.973), and the aril weight and the juice content of pomegranate fruits (R^2, 0.994). The results of the present study made it possible to estimate the number of arils, aril weight, and juice content of Wonderful cultivar pomegranate fruits by its fruit weight at harvest or after 90 days of storage with or without modified atmosphere packaging. The results also showed that the final fruit size of pomegranate fruits is not only related with the number of arils but also related with aril sizes. Understanding this phenomenon has important implications on cultural practices. The number of arils is mainly related with pollination and fertilization which occur during early fruit set and fruit development. Therefore, careful plant management is important during this time of period for obtaining larger fruits. But it must be kept in mind that the final fruit size is not determined at that time and regular fertilization is also crucial to enlarge arils.

The important result of the present study is that the modified atmosphere packaging has a clear effect on the protection of fruit weight, aril weight, and juice content. The loss in the fruit weight after 90 days of storage is found to be between 6.70 and 14.28% without MAP and between 2.64 and 6.24% with MAP. Similar results have been determined for aril weight and juice content. Last but not least, an important finding of the present study indicated that the larger fruits show higher weight loss.

Author details

Hatice Serdar* and Serhat Usanmaz

*Address all correspondence to: hatice.serdar.1@gmail.com

European University of Lefke, Lefke, Cyprus

References

[1] Lye C and Rural Industries Research and Development Corporation of Australian Government. Pomegranate: Preliminary assessment of the potential for an Australian industry [Internet]. 2008. Available from: https://rirdc.infoservices.com.au/downloads/08-153.pdf [Accessed: 04.08.16]

[2] Kahramanoğlu İ and Usanmaz S. Pomegranate Production and Marketing. 1st ed. Boca Raton, USA: CRC Press, Taylor & Francis Group; 2016. 148 p. ISBN: 9781498768504

[3] Morton J. Pomegranate. In: JF Morton, editor. Fruits of Warm Climates. 1st ed. Miami, FL: Florida Flair Books; 1987. pp. 352-355. ISBN: 9780961018412

[4]   Holland D, Hatib K and Bar-Ya'akov I. Pomegranate: Botany, horticulture, breeding. In: J Janick, editor. Horticultural Reviews. 35th ed. Hoboken, NJ, USA: John Wiley & Sons, Inc.; 2009. pp. 127-191. DOI: 10.1002/9780470593776.ch2

[5]   Adams F. Genuine Works of Hippocrates. 1st ed. New York, USA: William Wood and Co.; 1849. 466 p.

[6]   Jurenka J. Therapeutic applications of pomegranate: A review. Alternative Medicine Review. 2008;**13**(2):128-144.

[7]   Caleb OJ, Opara UL and Witthuhn CR. Modified atmosphere packaging of pomegranate fruit and arils. Food and Bioprocess Technology . 2012;**5**(1):15-30.

[8]   Al-Mughrabi MA, Bacha MA and Abdelrahman AO. Effects of storage temperature and duration on fruit quality of three pomegranate cultivars. Journal of King Saud University. 1995;**7**(2):239-248.

[9]   Elyatem SM and Kader AA. Post-harvest physiology and storage behaviour of pomegranate fruits. Scientia Horticulturae. 1984;**24**:287-298. DOI: 10.1016/0304-4238 (84)90113-4

[10] Opara UL, Mahdoury AA, Al-Ani M, Al-Khanjari SA, Al-Yahyai R, Al-Kindi H, Al-Khanjari SS. Physiological responses and changes in postharvest quality attributes of 'Helow' pomegranate variety (*Punica granatum* L.) during refrigerated storage. In: International Conference of Agricultural Engineering; 31 August–4 September 2008; At Iguassu Falls City, Brazil. Bonn, Germany: International Commission of Agricultural Engineering (CIGR), Institut fur Landtechnik; 2008.

[11] Kazemi F, Jafararpoor M and Golparvar A. Effects of sodium and calcium treatments on the shelf life and quality of pomegranate. International Journal of Farming and Allied Sciences. 2013;**2**(S2):1375-1378.

[12] Farber JN, Harris LJ, Parish ME, Beuchat LR, Suslow TV, Gorney JR, Garrett EH and Busta FF. Microbiological safety of controlled and modified atmosphere packaging of fresh and fresh-cut produce. Comprehensive Review in Food Science and Food Safety. 2003;**2**:142-160. DOI: 10.1111/j.1541-4337.2003.tb00032.x

[13] Fawole OA and Opara UL. Effects of storage temperature and duration on physiological responses of pomegranate fruit. Industrial Crops and Products. 2013;**47**(1):300-309. DOI: 10.1016/j.indcrop.2013.03.028

[14] Bayram E, Dündar Ö and Özkaya OK. The effects of different packaging types on the cold storage 'Hicaznar' pomegranates (second year). In: M Erkan and U Aksoy, editors. X International Controlled and Modified Atmosphere Research Conference; 4-7 April 2009; Antalya, Turkey. ISHS Acta Horticulturae; 2010. DOI: 10.17660/ActaHortic.2010.876.25

[15] Mirdehghan SH and Ghotbi F. Effects of salicylic acid, jasmonic acid and calcium chloride on reducing chilling injury of pomegranate (*Punica granatum* L.) Fruit. Journal of Agricultural Sciences and Technologies. 2014;**16**(1):163-173.

[16] Akbarpour A, Hemmati K and Sharifani M. Physical and chemical properties of pomegranate (*Punica granatum* L.) fruit in maturation stage. American-Eurasian Journal Agricultural & Environmental Sciences. 2009;**6**(4):411-416.

[17] Polat AA, Çalışkan O and Kamiloğlu O. Determination of pomological characteristics of some pomegranate cultivars in Dörtyol (Turkey) conditions. In: A D'Onghia, I Batlle, T Caruso, L Corelli Grappadelli, O Callesen, J Bonany, editors. XXVIII International Horticultural Congress on Science and Horticulture for People (IHC2010): International Symposium on the Challenge for a Sustainable Production, Protection and Consumption of Mediterranean Fruits and Nuts; Lisbon, Portugal. ISHS Acta Horticulturae; 2012.

[18] Wetzstein HY, Zhang Z, Ravid N and Wetzstein ME. Characterization of attributes related to fruit size in pomegranate. HortScience. 2011;**46**(6):908-912.

[19] Tabatabaekoloor R and Ebrahimpor R. Effect of storage conditions on the postharvest physico-mechanical parameters of pomegranate (*Punica granatum* L.). Asian Journal of Science and Technology. 2013;**4**(5):82-85.

[20] Arendse E, Fawole OA, Opara UL. Effects of postharvest storage conditions on phyto-chemical and radical-scavenging activity of pomegranate fruit (cv. Wonderful).. Scientia Horticulturae. 2014;**169**(1):125-129. DOI: 10.1016/j.scienta.2014.02.012

[21] Oğuz Hİ, Şen F and Eroğul D. Güneydoğu Anadolu bölgesinde farklı lokasyonlarda yetiştirilen 'Katırbaşı' nar (*Punica granatum* L.) çeşidinin depolanma süresince bazı fiziksel ve biyokimyasal içeriklerindeki değişimlerin belirlenmesi. YYÜ Tarım Bilimleri Dergisi. 2014;**24**(3):309-316.

6

Processing and Preservation of Fresh-Cut Fruit and Vegetable Products

Afam I.O. Jideani, Tonna A. Anyasi,
Godwin R.A. Mchau, Elohor O. Udoro and
Oluwatoyin O. Onipe

Additional information is available at the end of the chapter

Abstract

Fruits and vegetables are plant derived products which can be consumed in its raw form without undergoing processing or conversion. Fresh-cut fruits and vegetables (FFV) are products that have been cleaned, peeled, sliced, cubed or prepared for convenience or ready-to-eat consumption but remains in a living and respiring physiological condition. Methods of preserving FFV to retain its wholesomeness includes washing with hypochlorite, hydrogen peroxide, organic acids, warm water and ozone for disinfestation and sanitization; use of antimicrobial edible films and coatings; and controlled atmosphere storage and modified atmosphere packaging of fruits and vegetables. Exposure of intact or FFV to abiotic stress and some processing methods, induces biosynthesis of phenolic compounds and antioxidant capacity of the produce. Conversely, loss of vitamins and other nutrients has been reported during processing and storage of FFV, hence the need for appropriate processing techniques to retain their nutritional and organoleptic properties. FFV are still faced with the challenge of quality retention and shelf life preservation mostly during transportation and handling, without impacting on the microbiological safety of the product. Hence, food processors are continually investigating processes of retaining the nutritional, organoleptic and shelf stability of FFV.

Keywords: fresh-cut fruits, fresh-cut vegetables, preservation, bioactive compounds, organoleptic properties

1. Introduction

According to the International Fresh-Cut Produce Association (IFPA), fresh-cut fruit and vegetable products (FFVP) are defined as fruits or vegetables that have been trimmed, peeled or cut into a 100% usable product which has been packaged to offer consumers high nutrition and flavour while still maintaining its freshness [1, 2]. The importance of fresh-cut produce lies in its major characteristics of freshness, convenience, nutrient retention and sensory quality while providing extended shelf life [3, 4]. Fresh-cut fruits and vegetables (FFV) are products partially prepared and which require no additional preparation for their use. This makes it unavoidable that their overall quality diminishes during processing and storage. It is made more so, as the operations involved in preparing fresh-cut products damage the integrity of the cells, promotes contact between enzymes and substrates, increases the entry of microorganisms and creates stress conditions on the fresh-cut produce [4, 5].

According to Artes-Hernandez et al. [6] FFVP are also referred to as products prepared with slight peeling, cutting, shredding, trimming and sanitizing operations and which have been packed under semipermeable films and stored under refrigerated temperature. Fresh-cut products are also reported to contain similar nutrients and ingredients as whole products with the added advantage of short time preparation and the low prices at which they are been sold [7]. Fresh-cut products constitute a major rapidly growing food segment which is of interest to food processors and consumers. The fresh-cut industry is expanding more rapidly than other sectors of the fruit and vegetable market due to its supply of both the food service industry, retail outlet as well as its expanding production and access to new markets across the globe. The growth rate of the sector is reported to be in the region of billions of dollars in recent years with USA as the main producer and consumer while the UK and France follows after [6]. FFVP presently sold in markets across the globe includes: lettuce (cleaned, chopped, shredded), spinach/leafy greens (washed and trimmed), broccoli and cauliflower (florets), cabbage (shredded), carrots (baby, sticks, shredded), celery (sticks), onions (whole peeled, sliced, diced), potatoes and other roots (peeled, sliced), mushrooms (sliced), jicama/zucchini/cucumber (sliced, diced), garlic (fresh peeled, sliced) as well as tomato and pepper (sliced) [2].

Despite the fact that food processing methods extend the shelf life of fruit and vegetable products, processing of fresh-cut produce however reduces the shelf life of the commodity, rendering the product highly perishable as a result [6]. This biological changes may lead to flavour loss, cut-surface discolouration, decay, rapid softening, increased rate of vitamin loss, shrinkage as well as shorter shelf life of the fresh-cut produce. Interactions between intracellular and intercellular enzymes with substrates as well as increased water activity may also lead to flavour and textural changes upon processing [8]. A major effect of fresh-cut processing is stress on vegetable tissues with the resultant phytochemical accumulation and loss induced through reduced activity in key enzymes of secondary metabolic pathways. Fresh-cut processing also results in cell breakdown as well as the release of intracellular products such as oxidizing enzymes thereby quickening product decay [2].

Several factors are reported to affect the overall quality of fresh-cut produce. Among many of such factors is appearance [1, 9]. Appearance according to Kays [10] and Lante and Nicoletto [11] in combination with size, shape, form, colour as well as the absence of defects are factors which greatly affects the purchase of fresh-cut produce by consumers. All of these factors can also be influenced by several pre-harvest factors. Available nutrients inherent in fruits and made available upon consumption includes antioxidant vitamins beta-carotene (pro-vitamin A), α-tocopherol (vitamin E) and ascorbic acid (vitamin C). Research has also shown that regular consumption of fruits and vegetables reduces risk of cancers, cardiovascular diseases and several inflammations [12, 13]. This apart from regular body exercise and genetics has made fruit and vegetable consumption one of the main factors that contributes to a healthy lifestyle. With studies showing the nutritional benefits of fruits and vegetables, consumption of FFVP therefore promotes health through increase in the supply of antioxidant and other phytochemical nutrients to the body.

2. Processing of FFV

FFV have been known to have a shorter shelf-life compared to intact fruit and vegetable products due mainly to processing. Several processes involved in the production of FFVP have been known to alter greatly the shelf stability of the cut-produce. There are traditional processing procedures involved in obtaining fresh-cut products and this procedures usually requires an order of unit operations such as peeling, trimming, shredding, dicing, cutting, washing/disinfecting, drying and packaging. Shelf life extension of the cut produce is therefore dependent on a combination of these unit operations as well as proper temperature management during storage, use of antibrowning agents, proper packing conditions as well as good manufacturing and handling practices [6, 7]. The unit operations required in the handling and processing of FFVP is shown in **Figure 1**.

Cutting

An essential aspect of processing of fresh-cut produce is cutting. Cutting helps divide whole harvested fruit and vegetable products into minute fractions before packaging. The effect of cutting however on the products is the wounding stress which the cut tissues are allowed to suffer thus accelerating the rate of spoilage and deterioration of the cut produce [14]. Cutting has been attributed to be the main factor responsible for the deterioration of FFV thereby enabling the product to experience a more rapid rate of deterioration than whole products [15]. Cutting increases respiration rate [16], induces deteriorative changes associated with plant tissue senescence and thus the consequential decrease in shelf life when compared to the unprocessed produce [4]. Cutting shape as well as the sharpness of the cutting blade has been attributed as some factors that affect the quality attributes of fresh-cut products [17, 18]. The works of Portela and Cantwell [19] showed that melon cylinders cut with a blunt blade demonstrated higher ethanol concentrations, off-odour scores, electrolyte leakage, and increased potential for ethylene secretion when compared to products processed with a sharp blade. It was also reported that use of sharp cutting implements reduces wound response, lignin accumulation, white blush, softening and microbial growth in fresh-cut carrots [18, 20–22]. Cutting-induced injury has been implicated as affecting the immediate visual quality of fresh-cut products and has also been known to have longer-term effects on metabolism with the concomitant quality changes that are detected at a later time. The actual cutting process

RECOMMENDED MAX. TEMPERATURE		GENERAL STEPS. UNIT OPERATIONS
	<20ºC	HARVESTING
	<10ºC	TRANSPORTATION
DIRTY AREA	2-4ºC	PRECOOLING AND STORAGE
	8ºC	SORTING AND CLASSIFICATION
	8ºC	WHOLE PRODUCT WASHING
	0-2ºC	COOLING
	10-12ºC	CONDITIONING: PEELING, CUTTING, GRATTING, SHDREDDING...
CLEAN AREA	0-2ºC	PREWASHING, WASHING & DISINFECTION
	0-2ºC	RINSING
	8ºC	DEWATERING AND CENTRIFUGATION
	<5ºC	WEIGHTING AND MIXING
	<5ºC	MODIFIED ATMOSPHERE PACKAGING (ACTIVE OR PASIVE)
	<5ºC	QUALITY CONTROL AND BOXING
	<5ºC	LOAD PREPARING, TRANSPORTATION AND DISTRIBUTION
	5ºC	RETAIL SALE
	5ºC	CONSUMER

(Processing Line / FACTORY)

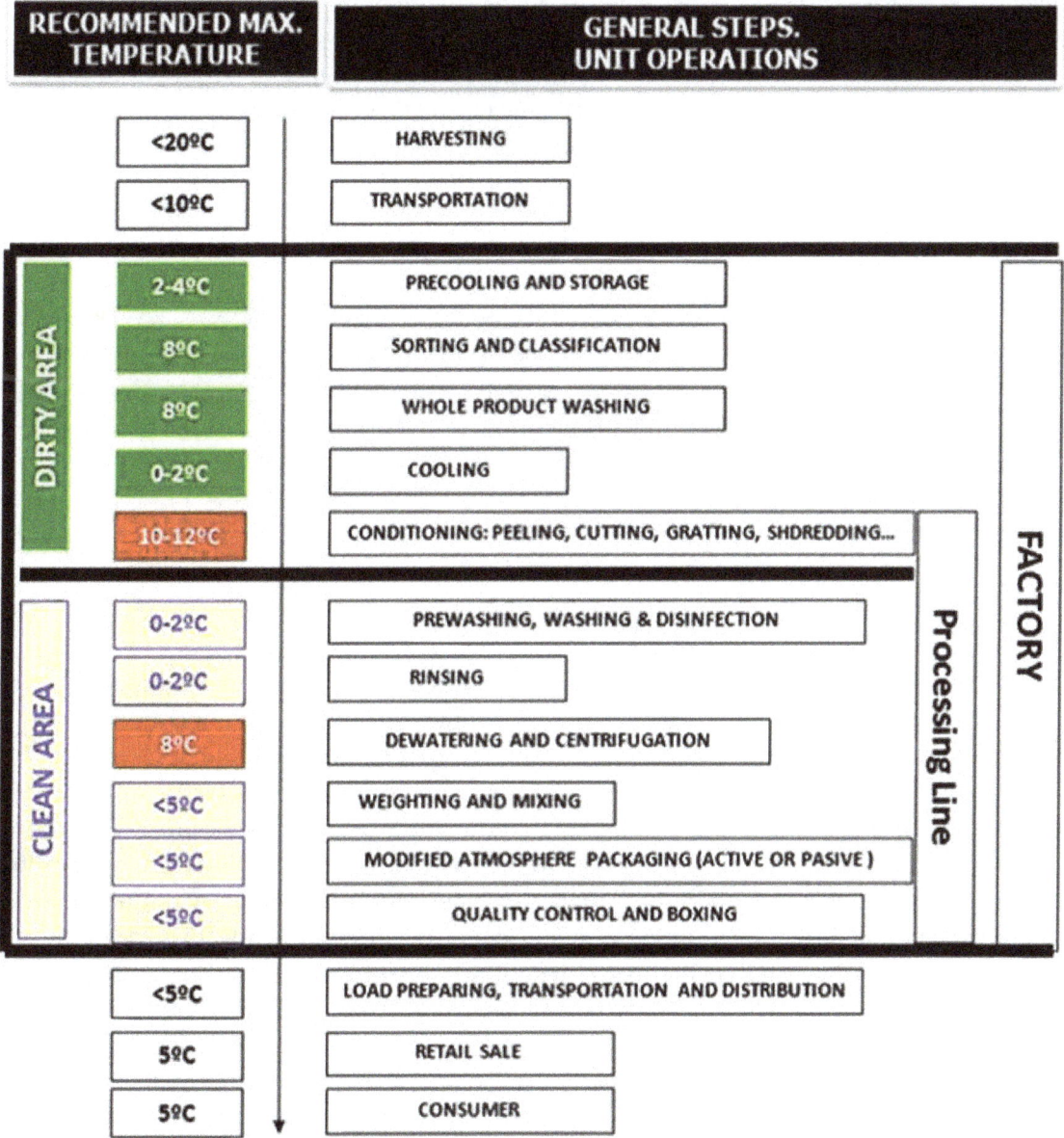

Figure 1. Unit operations and maximum recommended temperatures for each step in the processing of fresh-cut fruit and vegetable produce. Adapted from Artes-Hernandez et al. [6].

results in great tissue disruption as formerly sequestered enzymes and substrates mix are found to mix, hydrolytic enzymes released, while signalling-induced wounding responses may be initiated [23].

During the process of cutting, phenolic metabolism takes place: breakage of the plasma membrane with the resultant effect of inducing oxidative enzymatic reactions thus triggering browning of tissues and oxidation of polyphenols [14, 24]; and production of injury signals which induces the secretion of more secondary metabolites including phenolic antioxidants to heal the wound damage [14, 25, 26]. It has been reported that the content of phenolic acids increases in fresh-cut products. This fact can be attributed to the cutting of fresh fruits and vegetables with knives thereby inducing the activity of polyphenol oxidase (PPO) in the cut fresh fruit and vegetables. FFV are thus easily susceptible to browning reaction as a result [27, 28]. Accumulation of phytochemicals can also be as a result of altered O_2 and CO_2 levels during packaging as well as the use of preservatives such as [14] ascorbic and citric acid [28–31].

Wounding as a result of cutting has been attributed as one of the basic source of stress experienced by fresh-cut produce. Some factors can however affect the wound response of the fresh-cut produce and these factors include stage of maturity, cultivar, storage, processing temperature, cutting method, water vapour pressure as well as O_2 and CO_2 levels [18, 19]. According to literature, wounding stress as a result of cutting of fruit and vegetables has been shown to increase the antioxidant activity as well as the polyphenolic content in fresh-cut produce such as carrots [32, 33], celery [34], lettuce [35], broccoli [36], mushroom, onions, and mangoes [37]. Consequences of wounding includes increase in respiration rate; production of ethylene; oxidative browning; water loss; and degradation of membrane lipids [4, 5]. This therefore increases the susceptibility of FFV to increased perishability than their source commodity [38].

Sanitation and hygiene in processing facilities

During the production process, cut fruits are exposed to environmental microbes in the processing facility. Reducing the level and rate of contamination will be dependent on the use of the appropriate disinfectants and sanitizers. One of such disinfectants of high use in the FFV industry is chlorine. The use of chlorine as a disinfectant is however of great concern and is presently prohibited in some European countries due to issues of public health [39]. Chlorine is generally used in the food industry due mostly to its low price and its wide application of antimicrobial effectiveness [39, 40]. However, under certain conditions, chlorine has been shown to be weak in reducing microbial loads [41] as it can easily be inactivated by organic matter [40, 42] and its action is highly pH reliant. Chlorine has also been shown to produce unhealthy by-products which are carcinogenic and mutagenic such as chloroform, trihalomethanes, chloramines and haloacetic acids, when reacting with organic molecules [43, 44]. Chlorine is also corrosive with its use banned in some European countries such as Belgium, Denmark, Germany and The Netherlands [40, 45–47]. Presently, alternative chemical compounds, biological methods and physical technologies which are more environmentally friendly and possess less risk to the health of workers and consumers have been developed to replace the use of chlorine [45–52].

Concentrations of 50–200 ppm and exposure time of 5 min of chlorine is commonly applied as hypochlorous acid and hypochlorite and as disinfectant in the FFV industry in order to enhance microbial safety of the produce [1, 49]. The exposure time of 5 min (depending on the microorganism) has been shown in literature as the maximum exposure time required as longer times of > 5–30 min did not result in increased removal of the pathogenic organisms [39, 53].

In the handling and processing of FFV, common practices are undertaken and needs to be taken note of. These practices consist of protection from damage as a result of poor handling and poor functioning of machinery, foreign body contamination, improper washing, drying and unhygienic practices by personnel. Hence worker sanitation which is most often neglected in the fresh cut industry in collaboration with good manufacturing practices must be enforced by food processors. In accompanying this process, training of food handlers in food hygiene techniques must be undertaken [6].

Presently, new and alternative technologies for safety, improved quality and extended shelf life of processed fresh-cut products have been developed. Such technologies include: ozone (O_3), a strong oxidizing agent in destroying microorganisms which has also been suggested as an alternative to sanitizers due to its effectiveness at low concentrations, short contact times and in the breakdown of nontoxic products; chlorine dioxide (ClO_2), which is known for its efficacy against pathogenic spores, bacteria and viruses; organic acids and calcium (Ca) salts applied for maintenance of cell wall structure and firmness (Ca), inhibition of enzymatic and non-enzymatic browning as well as in the prevention of microbial growth at heights that did not affect flavour of the fresh-cut products with their efficacy against microbes higher for bacteria than molds; electrolyzed water employed due to its strong bactericidal effect against pathogens and spoilage microbes [6].

2.1. Inhibition of browning in FFV

It has been shown that fresh-cut process increases the metabolic activity mainly as a result of the enzymes polyphenol oxidase (PPO) causing discoloration and peroxidase (POD) causing enzymatic browning as well as de-compartmentalization of enzymes and substrates in tissues causing changes in flesh colour [54]. PPO can induce the browning occurrence by catalyzing the oxidation of phenol to o-quinones which are polymerized to produce brown pigments. Postharvest techniques maintaining the quality of fresh-cut fruit have been investigated by several researchers [55–57] including physical and chemical treatments. Many anti-browning agents or mixtures have been investigated like: calcium ascorbate with citric acid and N-acetyl-L-cysteine [58], citric acid [59], ascorbic acid with citric acid and calcium chloride [30], 4-hexylresorcinol with potassium sorbate and D-isoascorbic acid [60] and modelling of the effects anti-browning agents on colour change in fresh-cut [57].

In a study using mathematical modelling (**Table 1**), the effects of different anti-browning compounds (ascorbic acid, citric acid, L-cysteine and glutathione) at four concentrations of 0% as control, 0.5, 1.5 and 2.5% on L^*-value, hue angle, brown scores and brown pigments of fresh-cut mangoes were investigated by Techavuthiporn and Boonyaritthonghai [57] and they observed similar changing tendency of L^*-value and hue angle decreasing in time during storage at 10°C, while the brown scores and amount of brown pigment increased. They also observed that treatment with L-cysteine or glutathione was effective in suppressing tissue metabolism, PPO and POD activities, while citric acid significantly inhibited the growth of microorganisms.

2.2. Microbial safety of FFV

The unit operation employed in processing of FFV involves peeling, cutting, slicing and shredding; all of which cause disruption of surface cells, tissue and cytoplasm exposure, coupled with high water activity and low pH; thereby providing a breeding ground for growth of pathogenic

Treatments	K_L ± S.E. (day^{-1})[a]	K_H ± S.E. (day^{-1})[a]	K_{BS} ± S.E. (day^{-1})[a]	K_{BP} ± S.E. (day^{-1})[a]
Control	0.0209 ± 0.0012	0.0158 ± 0.0008	0.2604 ± 0.0474	0.2616 ± 0.0187
Ascorbic acid				
0.5%	0.0132 ± 0.0012	0.0132 ± 0.0002	0.3313 ± 0.04567	0.1566 ± 0.0231
1.5%	0.0088 ± 0.0007	0.0090 ± 0.0004	0.2146 ± 0.0777	0.1272 ± 0.0062
2.5%	0.0066 ± 0.0010	0.0079 ± 0.0003	0.2231 ± 0.0572	0.1186 ± 0.0129
Citric acid				
0.5%	0.0213 ± 0.0023	0.0120 ± 0.0014	0.1859 ± 0.0466	0.1981 ± 0.0212
1.5%	0.0210 ± 0.0030	0.0098 ± 0.0012	0.1642 ± 0.0971	0.1559 ± 0.0177
2.5%	0.0164 ± 0.0023	0.0097 ± 0.0007	0.1443 ± 0.1053	0.1386 ± 0.0062
L-Cysteine				
0.5%	0.0068 ± 0.0019	0.0095 ± 0.0011	0.1776 ± 0.0521	0.1427 ± 0.0263
1.5%	0.0053 ± 0.0008	0.0077 ± 0.0011	0.1342 ± 0.0400	0.1188 ± 0.0127
2.5%	0.0025 ± 0.0002	0.0041 ± 0.0001	0.1385 ± 0.0213	0.0653 ± 0.0047
Glutathione				
0.5%	0.0180 ± 0.0010	0.0107 ± 0.0004	0.1866 ± 0.0487	0.1993 ± 0.0095
1.5%	0.0094 ± 0.0011	0.0089 ± 0.0004	0.1676 ± 0.0262	0.1311 ± 0.0115
2.5%	0.0061 ± 0.0007	0.0087 ± 0.0004	0.1329 ± 0.0229	0.0973 ± 0.0167

[a]k_L and k_H represent the estimated rate constant of decreasing L^*-value and Hue angle in $Q_t = Q_0 \cdot e^{-kt}$ (Eq. 1) and k_{BS} and k_{BP} represents the estimated rate constant of increasing brown colour (score) and brown pigment in $Q_t = Q_0 \cdot e^{kt}$ (Eq. 2). Q_t = measured value of colour variables at time t, Q_0 = initial value of colour variables at time zero, t the storage time (day), and k reaction rate constant (day^{-1}).
Adapted from Techavuthiporn and Boonyaritthonghai [57].

Table 1. The estimated parameters k and the standard error of estimates (S.E.) in Eqs. (1) and (2) for dipped fresh-cut mangoes at different concentration and anti-browning agents.

microorganisms such as *Escherichia coli* 0157H7, Salmonella spp. and *Listeria monocytogenes*. Peeling and cutting of fruits and vegetables removes the protective epidermal layer, thus exposing the product to air and possible contamination by bacteria, molds and yeast. Contamination of produce can occur at any stage from production till consumption. During growth, harvest, transportation and further processing and handling, fresh-cut produce can be contaminated with pathogens from human, animal, or environmental sources [40, 61]. Since most FFV are consumed raw without any further treatment, consumption poses a potential health risk in cases where there is contamination. Several outbreak incidences have been documented by Centre for Disease Control and Prevention as reviewed by Ramos et al. [40]. Some of the sources of these microorganisms include soil, manure, silage, sewage, water, raw meat, and domestic animals. Presence of up to 10^4 of *Listeria monocytogenes* cells can cause food infection with up to 90 days incubation period. Symptoms of its infection includes flu-like symptoms, septicemia, encephalitis, still birth and abortion in pregnant women [62]. There are a number of factors affecting the microbial safety of FFV and which will be elucidated as follows:

Product

Microbial growth and survival depend largely of the quality or type of fruit or vegetable in question. The quality factors of a product that may affect microbial growth include pH, water activity, respiration rate, type of packaging, competitive microflora and innate antimicrobials [61]. The pH of a product strongly influences microbial growth. Most vegetables are known to have a pH of ≥ 5 which supports the growth of pathogenic microorganisms. Despite the acidic pH of most fruits, organisms such as *E. coli* 0157H7 and some *Salmonella* spp., still grow and survive in such acidic environment. *L. monocytogenes* was reported survive on cantaloupe & melon cubes [63] and apple slices at pH 3.42 [64]; Salmonella also survived on apple slices with 4.1 pH [65]. Fresh fruits and vegetables provide an ecological niche to some microorganisms; and these vary from one product to another depending on the type of product, climatic conditions, geographical location, harvesting, handling and transportation. Microflora of fruits and vegetables include bacteria such as *Erwinia herbicola*, *Enterobacter agglomerans*, *Xanthomonas*, *Leuconostoc mesenteroides*, *Lactobacillus* spp. *Flavobacterium*; and moulds like *Penicillium*, *Fusarium*, and *Aspergillus* [40]. An antagonistic behaviour of native microflora of fruits and vegetables against pathogenic microorganisms have been reported [61]. The growth of *L. monocytogenes* on lettuce was reduced by *Enterobacter* [66]. The natural microflora may help control growth of pathogenic microbes through (a) direct competition for nutrients and space; (b) production of antimicrobials [61].

Processing

Peeling and cutting increases respiration and ethylene secretion rate of fruits which in turn increases senescence, which makes more sugar available thereby allowing rapid microbial growth on fresh-cut fruits. Microbial contamination of fresh-cut produce can also be facilitated by cross-contamination of produce through: (a) transfer of organisms from surface of fruits onto FFV; (b) attachment of pathogens onto shredders and slicers which can be re-introduced, for instance L. monocytogenes has been recovered from the environment of vegetable processing plant [61]; (c) re-use of the same water for washing fruits and vegetables allows transfer of microorganisms from contaminated parts to uncontaminated parts. Packaging conditions also has influence on the growth of pathogenic microorganisms during storage. Modified atmosphere packaging which uses low oxygen and increased carbon dioxide for preservation of fresh-cut produce influences growth of pathogenic organisms. *E. coli* 0157:H7 grew on chicory slices when atmospheric CO_2 was increased to 30% when stored at 13 or 20°C. Storage temperature is one of the most important factor that affects the growth and survival of pathogenic microorganisms. Storage of FFV at temperatures ≤ 4°C reduces the growth of psychotropic organisms such as L. monocytogenes which can survive at low temperatures. However coliforms like Salmonella and *E. coli* 0157:H7 are unable to survive at low temperatures, but they proliferate at ambient temperatures. Fluctuations in storage temperatures should be avoided at all costs in order to maintain the safety of fresh-cut vegetables from microbial contamination.

2.2.1. Methods for detection of spoilage and pathogenic microorganisms on fruits and vegetables

2.2.1.1. Isolation using plate count method

Conventional method of bacterial enumeration works exquisitely on the recovery of viable bacteria from fruits and vegetables. Many food microbiology laboratories lack availability and utilization of novel molecular-based technologies. Due to this, conventional methods such as standard plate count, selective or differential media for isolation and detection and of bacteria samples; and further identification using biochemical methods such as gram stain and other commercially available profiling systems [67]. Non-selective media such as nutrient agar, tryptone soy agar can aid proliferation of bacteria cells at incubation time of 4 h before subsequent transfer to a selective media such as Mac Conkey agar for *E. coli*, and Salmonella-Shigella agar for *Salmonella* detection. Bacteria can be further separated into gram positive and negative using gram stain. The process of transfer from non-selective to selective media allows sub-lethally injured bacteria to recover in time for detection. Other biochemical tests used as follow up to conventional isolation of pathogenic bacteria include catalase test for identification of gram positive bacteria and oxidase test for gram negative bacteria [67].

2.2.1.2. Use of fluorogenic and chromogenic media

Another media used for isolation and detection of pathogenic bacteria is the use of fluorogenic and chromogenic culture media. Identification of bacteria on fluorogenic media is made possible through the incorporation of enzyme substrates (which consists of a sugar/amino acid-fluorogen complex) into a selective media which in turn speeds up biochemical confirmation of the bacterial identity [68]. For example, methylumbelliferyl is a fluorogen that has been incorporated in an array of media for detection of coliforms such as *E. coli* 0157:H7 which forms a blue fluorescence upon exposure to ultraviolet light. Rainbow agar, BCM O157:H7 (and other coliforms), CHROMagar, and Colilert are some examples of fluorogenic or chromogenic media for coliforms detection. The most commonly used culture reference methods for the detection of *Listeria* in foods are the ISO 11290 standards [69, 70]; FDA-BAM method to isolate *Listeria* spp. from vegetables [71].

2.2.1.3. Molecular-based methods

Nucleic acid-based systems designed for the detection of genomic DNA specific to particular microorganisms are capable of achieving rapidly and highly sensitive identification even when the target microbe is present in low numbers [72]. In order to achieve this, polymerase chain reaction (PCR) is quite useful. PCR is a molecular-based detection method. This method is focused on extraction of bacteria DNA; and it works best when there is enough bacterial cells from which is boosted by the enrichment step. Enzyme-linked immunosorbent assay (ELISA) is another molecular-based method that works on the principle of antigen-antibody interaction. It is more sensitive and specific to a bacteria strain than standard plate count method. Steps in Antigen (food slurry/extract) is added to sample wells in a microtiter plate containing an antibody with specificity to the target molecule. ELISA has been successfully used to detect virulence determinants of pathogens such as *Campylobacter*, *E. coli* 0157:H7 and *L. monocytogenes*. Serotypes of *Listeria* spp. were categorized based on specific heat-stable somatic (O) and heat-labile flagellar (H) antigens [71]. ELISA procedure entails adding sample, washing, adding antibody complexes, adding detection reagents; and these steps are labour intensive. This has led to automation of the ELISA process, thus cutting back on time and labour. A good example is BioMerieux's Vidas System in which the entire procedure is finished in 2 h after addition of overnight enrichment broth. This system is available for assays of pathogenic bacteria such as *L. monocytogenes*, *Salmonella*, *E. coli*, and *Campylobacter* [72].

2.2.1.4. Rapid methods of microbial pathogenic detection

Standard methods of isolation and identification of pathogenic microorganisms in foods are slower and time consuming which has led to demand of quicker methods; and for the past two decades the latter have been developed for both on-site and laboratory tests. A method can be characterized as rapid when it gives quicker results that the conventional method. Other factors that determine its effectiveness are sensitivity, standardization, reliability, accuracy, specificity, evaluation, ease of use, cost, validation, convenience and potential for automation. Most of the advances in development of rapid methods are in molecular-based methods and other areas including impedance and conductance, bacteriophages, biosensors, microscopy as well as in miniaturized, automated biochemical detection kits [72]. There are currently many diagnostic systems like RapID, Minitek, API, Biolog, MicroID, Crystal ID and VITEK systems which are commercially available for identifying different microorganisms [8].

3. Prevention of contamination, safety and hygiene practices during processing of FFV

The rate of contamination of FFVP after processing (cutting or wounding) is greater when compared to those of whole fruits and vegetables [73, 74]. This has been largely attributed to high moisture content in the fruits and vegetables as well as wound occurring in the tissues due to processing [75]. Wounding of tissue as a result of slicing, cutting or peeling releases the nutrients inherent in these products thereby enhancing microbial contamination and growth [76]. Upon growth of these microbes on the FFV surface, susceptibility to the formation of biofilms increases in the produce thus bringing about difficulty in the elimination of these microbes [77]. Microbial biofilms are thus able to attach, grow and spread to any surface with the cells associated with the biofilms possessing an advantage in growth and survival over planktonic cells. Growth and survival advantage over planktonic cells by biofilms has been attributed to the formation of exopolysaccharide matrix which surrounds the biofilms, thereby building a wall against the environment and protecting the biofilms from sanitizers [74, 78].

Pathogens of major concern in FFVP include *Listeria monocytogenes*, pathogenic *Escherichia coli* and *Salmonella* spp. A number of human pathogens have also been implicated in the contamination of FFV with a reported increase in recent years in the number of produce-linked foodborne occurrences. Agricultural practices during harvesting, human handling, quality of water and soil, contaminated equipment, processing methods, use of contaminated packaging materials as well as transportation and distribution have all been implicated in the contamination of fresh-cut produce [6, 79]. Similarly, microbial adhesion on conveyor belts, containers and food contact surfaces used along the food chain has been shown to lead to the formation of biofilms [41, 80]. Microbial contamination has also been shown to lead to internalization of pathogens into the fresh-cut fruit and vegetable produce. According to Golberg et al. [81], *E. coli* and *Salmonella typhimurium* are capable of penetrating the leaves of iceberg lettuce. The works of Seo and Frank [82] also showed that *E. coli* O157:H7 can penetrate between 20–100 μm below the surface of lettuce leaves. Internalization of such pathogens has been reported to occur in the stomata, vasculature, cut edges and intercellular tissues [83] with elimination of such pathogens rather impossible thus hindering assurance of product safety and rendering processing and preservatory methods completely futile [83, 84].

Some of the biological methods employed in the reduction of pathogenic attack and spoilage of processed FFV include bacteriocins such as lactic acid bacteria which produces organic acids and bacteriocins that can act as antimicrobials [85]. For instance bacteriocin nisin is a natural preservative produced by *Lactococcus lactis* and is effective against mostly gram positive bacteria [86, 87]. Nisin acts on the cell membrane of the microbe thereby forming pores that result in cell death during the process [86, 88]. Other biological methods employed include the use of bacteriophages which has found application as disinfectants and preservatives. Other biological methods applied include the use of bacteriophages used in the destruction of bacteria. Bacteriophages are virus that infect bacteria thereby bringing about their death and destruction as a result. Its advantages in its application includes specificity in action; effectiveness [89]; availability and accessibility; and reduced effects on the organoleptic properties of the fresh-cut products [90].

Enzymes have also been employed in the control of pathogenic organisms in fresh-cut products. According to Simões et al. [91] and Thallinger et al. [92], enzymes are able to target directly the biofilms that interfere with their development process, speed up the formation of antimicrobials and even destroy mature biofilm. In attacking biofilms, enzymes mostly target the extracellular polymeric matrix which surrounds the biofilm cells and influences the shape of the biofilm structure as well as its resistance to shear forces [93]. Hence, enzymes can be used as an alternative to conventional chemical disinfectants in the removal of biofilms from fruits, leaves and other abiotic surfaces [39]. However, the use of enzymes requires prolonged contact times for effectiveness against biofilms and the fact that extracellular polymeric substances produced by biofilms are mostly heterogeneous confers some disadvantages on the use of enzymes [39]. Accordingly, use of enzymes alone as a biological control against pathogens in fresh produce does not guarantee total removal of biofilms. Lequette et al. [93] and Augustin et al. [94] therefore suggested the use of enzymes in combination with other treatments especially with antimicrobial agents.

4. Storage of FFV

4.1. Modified atmosphere packaging (MAP) of fresh-cut produce

One of the various ways of processing fresh-cut fruits and vegetable is the use of modified atmosphere packaging (MAP) of the produce. The process of MAP helps in altering the gaseous composition within a food packaging system. MAP relies greatly on the interface between the rate of respiration of the produce and the transfer of gases through the packaging material without any further alteration to the initial gas composition [95–98]. MAP can either be passive: which involves generation of MAP in a packaging material by reliance wholly on the natural process of respiration of the packaged produce as well as the permeability of the packaging film material in bringing about the desired gas composition. MAP can also be active: involving the replacement of the gaseous composition in a packaged material through the introduction of gas scavengers or absorbers such as ethylene scavengers, oxygen and carbon (iv) oxide, thereby establishing the preferred gas mixture within the package [95, 98–100].

FFV have a short shelf life due to respiratory metabolism (**Figure 2**). Modified atmosphere packaging (MAP) has been used to reduce the rate of respiration and water loss leading to prolonged storage period. MAP comprising of low O_2 and elevated CO_2 atmospheres have been used to extend the shelf life and leading to high organoleptic characteristics of pear [101], apple [102], mango [55] and peach [103] fresh cut. The effects of low O_2 and CO_2-enriched atmospheres associated to different packaging, traditional and compostable, on shelf life of fresh-cut nectarine slices stored (1°C) for 7 days by Maghenzani et al. [104] showed that low permeability of the film has a positive influence on weight loss and firmness as the less permeable film allowed a greater water retention, which caused a lower weight loss. They observed that MAP, acting on the respiratory metabolism, reduced respiratory metabolism with positive effect on colour, total soluble solids, titratable acidity, firmness and PPO activity, though efficacy differed among two cultivars of the fruit. Biodegradable films performs better than polyethylene film as a packaging material.

4.2. Freezing of FFVP

Freezing is a widely known and applied preservation process of various foods which offers the advantage of producing high-quality nutritious foods with prolonged shelf life. Freezing has also been described as one of the best methods used in preserving foods such as fruits and vegetables. Freezing of FFV will reduce the problem of spoilage experienced by the fresh-cut commodities. However, there is a perception by consumers that freezing reduces and affects negatively the nutritional composition of the fruits [106, 107]. A point of comparison is based on the fact that fresh produce could maintain its keeping quality in the consumer's home for a number of days prior to consumption [108].

During freezing most of the liquid water constituent of the food materials is transformed into ice, thereby reducing water activity, which slows down the physical and biochemical changes

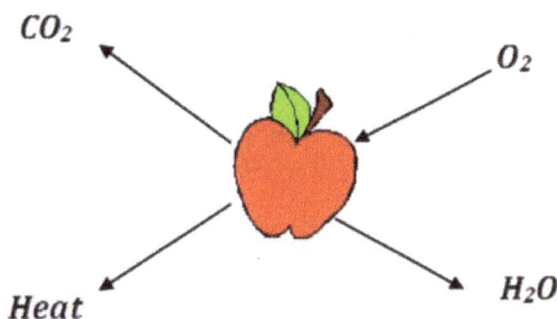

Figure 2. Fresh foods such as fruit and vegetables are alive and continue to respire after harvest. Reducing the respiration rate and reducing the heat produced through efficient airflow inside ventilated packaging is important in maintaining product quality [105].

involved in the deterioration of foods as well as the growth and reproduction of spoilage microorganisms. Fruits and vegetables are composed of approximately 85–90% water which crystallizes during freezing. The freezing process prevents microbial growth, reduces water activity, and decreases chemical and enzymatic reactions [109]. According to Jaiswal et al. [110], decrease in temperature experienced during freezing impedes metabolic reactions taking place in the fruit and vegetable after harvesting. Freezing also reduces the rate of microbiological activities occurring in the FFV positively affecting the overall product quality. During the process of freezing, conversion of water to ice brings about various stress mechanism such as volumetric change of water converting into ice, spatial distribution of ice within the system as well as the size of the ice crystal [111, 112]. The effect of this stress mechanism is the deterioration of the frozen products by affecting the texture and structure of the cut fruit and vegetable.

It is well known that FFV undergo faster physiological deterioration, biochemical changes and microbial degradation [113] which may result in degradation of the colour, texture and flavour. However, the high water content of fresh-cut products adversely affects the textural quality of the products after thawing due to the formation of ice crystals and water solids within the cell structure. When thawing takes place the cellular structure of the fruit and vegetables is destroyed [114]. The reduction in the product water content results in improved freezing performance and ameliorated product quality including better preservation of structural and textural characteristics [115]. Thus, in order to preserve the structural and textural characteristics and improve freezing performance, the water content of the fruits and vegetables are reduced by dehydration before freezing. Frozen fruits and vegetables are mostly consumed cooked with majority of vegetables blanched prior to freezing. Blanching action prior to freezing has been reported to influence greatly the structure of the vegetable thus resulting in an initial loss of firmness due to disruption of the plasma lemma and an increase in the ease of cell separation accompanied by swelling of the cell wall [112, 116].

Several novel freezing practices are presently being investigated to overcome the problems of FFV and other food produce undergoing physical and chemical changes as a result of freezing. One of such novel methods which is presently being explored is dehydrofreezing [115]. During dehydrofreezing process, the food is first dried up to a needed moisture content level before the onset of freezing. Hence it is aptly described as a process of freezing relatively dehydrated foods [117]. For fresh-cut products, non-thermal dehydration techniques such as vacuum and air drying are mildly applied prior to freezing. When the method of drying the FFV is through osmosis then it is termed osmodehydrofreezing. Dehydrofreezing is particularly well suited for fresh-cut fruits and vegetables due to the fact that reducing the moisture content in the produce will allow for the formation and expansion of ice crystals without damaging the cellular structure of the product [115]. Theoretically, the dehydration treatment not only reduces the amount of water to be frozen but also makes cell structures less susceptible to breakdown by changing cell turgor pressure [118]. The reduced water content has the potential to reduce the freezing time, the initial freezing point and amount of ice formed within the product [117]. As a consequence, the damage to plant cells caused by ice crystal formation and the post-thawing quality degradation such as softening and loss of good textural attributes are alleviated. Reduction in moisture content before freezing also leads to reduced freezing time since there is less water to freeze as well as a reduction in the amount of ice formed within the produce [119, 120]. According to Li and Sun [118], fruits and vegetables are said to exhibit better quality over those that are frozen without any form of reduction in moisture content.

Generally, the texture of a thawed frozen fresh-cut fruit and vegetable product is much softer than normal produce due to cell rupturing caused by expansion of the plant cells during freezing. Hence the recommendation that the moisture content of fresh-cut produce be reduced before freezing in order to mitigate the effect of freezing on the thawed product.

5. Nutritional content of FFV

FFV are derived from whole fruits by cutting them into desired shapes and sizes. However, peeling and cutting cause serious damage to vegetable tissues which leads to dissociation of cell components that brings about biochemical reactions such as accelerated oxidative browning and chlorophyll degradation. Other quality deterioration include water loss, development of off-flavours, stimulation of microbial growth and tissue softening which makes fresh-cut fruits have short shelf life [121–123]. Wounding stress as a result of cutting first causes the plasma membrane to break thereby inducing reaction of oxidative enzymes with existing phenolic compounds causing oxidation of the latter [24].

Nutritional value of fresh-cut fruits is usually a measure of vitamins A, B, C, E, polyphenolics and carotenoids; while that of vegetables include the previously mentioned vitamins, glucosinolates, carotenoids and polyphenolics through spectrophotometric and colorimetric methods [121]. Li et al. [123] tested the effect of cutting a whole pitaya fruit into slice, half and quarter slices on its nutritional quality. Their results revealed that the various cutting styles had little influence on vitamin C and soluble solids. However, total phenolic content, antioxidant activity, increased significantly with cutting wounding intensity up to first two days of storage before deterioration set in. Some nutritional contents of selected fruits and vegetables are highlighted in **Table 2**.

5.1. Sugars

Total sugar content of swede were not affected by storage temperature; while lower temperatures (0 & −2°C) increased the sugar content of turnip than higher temperatures as a result of glucose and fructose metabolism by enzymes at lower temperatures [124]. Benítez et al. [122] reported that soluble solid content of kiwi slices coated with aloe vera gel, alginate and chitosan did not significantly change up till day 8 of storage at 5°C.

5.2. Vitamin C

Vitamin C is the vitamin that usually degrades most rapidly and can be used as an index of freshness. Vitamin C is unstable in many vegetables such as asparagus [121]. There was no significant difference in the ascorbic acid contents of FC papaya stored at 10–20°C for 0 to 24 h

Fruits/veg	Sugars (°Brix)	Vitamin C (mg/100 g)	Polyphenols (mg/100 g)	Antioxidants (%DPPH)
Kiwi	12.6	44.99	-	-
Apple	-	3.00	329.30–660.00	12
Pitaya	0.15 g/kg	0.16 g/kg	0.6–1.2 g/kg	40–70
Papaya	11.82	13.3–18.0	1.9–3.5	-
Pineapple	-	76.27–142.41	42.28–46.70	30–60
Banana	-	46.75–53.25	60.25–85.24	70–90
Guava	-	65.16–198.25	96.51–178.51	-
Radicchio leaves	-	-	143–360	-
Swede	535.2 g/kg	-	-	-
Turnip	468.9 g/kg	-	-	-

Source [1, 122–124, 127].

Table 2. Nutritional content of some fruits and vegetables.

but a significant decrease was observed at storage temperature and time of 4C, 48 h [125]. There was no significant influence of cutting style on vitamin C content of pitaya fruit when cut in slice, half and quarter shape [123]. Exposure of fresh-cut banana, pineapple and guava slices to ozone for 0–30 min drastically reduced vitamin C contents of the fruits by 12.21, 46.44 and 67.13% respectively [126]. Vitamin C content of kiwi slices coated with chitosan and alginate, stored at 5°C depreciated at storage time from day 1 to day 11. However, kiwi slices coated with aloe vera gel significantly increased from 44.99 to 47.99 mg/100 g at the same storage conditions [122].

5.3. Polyphenols

When wounding stress occurs, plants produce injury signals to induce the production of more secondary metabolites including phenolic antioxidants to defense and heal the wounding damage [24]. Wounding stress also activates phenylalanine ammonia lyase (PAL) - an enzyme responsible for synthesis of phenolic compounds in plant tissues. For example, carrots synthesize lignin along wound barriers [127]. Flavonoid contents of fresh-cut papaya significantly increased at storage conditions of 20°C at 24, 36 and 48 h when treated with 405 nm LED illumination [125]. Li et al. [123] reported a gradual increase in total polyphenol content of pitaya fruit with storage time of 4 days. An approximate increase of 63, 78 and 90% was reported for slice, half and quarter slice respectively at day 2; after which a decline was observed. Total polyphenols and flavonoids in fresh-cut pineapple, and banana increased as the fruit slices were exposed to ozone for up to 20 min; but a reverse trend was observed for guava slices. The reason for increase polyphenols by ozone treatment was attributed to activation of PAL [126]. Upon 1 day in storage, total phenolic content of untreated and fresh-cut apples treated with citric acid and UV light decreased by 50% Ref. [128].

5.4. Antioxidants

Kim et al. [6] observed no significant difference in antioxidant capacity of fresh-cut papaya stored at 4–20°C for 0 to 48 h. Antioxidant capacity of fresh-cut pineapple and banana increased when exposed to ozone for 20 min and declined upon further treatment; while that of fresh-cut guava reduced with ozone treatment and increased when ozone exposure time was increased to 30 min [126]. A drastic loss of antioxidant potential of fresh-cut apples was observed in both untreated and fruit pieces treated with pulsed light and gellan-gum coatings during the first week of storage [129].

6. Emerging/future trends in processing and preservation of FFV

Fruits and vegetables remain important health food with low in fat, sodium and calories and high concentrations of vitamins, minerals and phytochemicals especially antioxidants protecting body cells against free radicals [130, 131]. Emerging technologies to fresh-cut fruits and vegetables to inactivate bacteria and viruses are focusing on modified atmosphere packaging process. The microbial inactivation effect of this technologies has to be further assessed. The number of studies is still low in the area of emerging technologies such as low-pressure application to reduce microbial populations in FFV. Very few studies have focused on viral inactivation during MAP processes. More evidence is needed that MAP process can contribute to reduce or eliminate specific foodborne pathogens to reduce the risk for foodborne infection associated with FFV when consumed as such or when used further in the food supply chain as ingredients.

Packaging material, including low density polyethylene (LDPE), laminated aluminium foil (LAF), high density polyethylene (HDPE), polypropylene (PP), polyethylene (PE), is an essential component of the FFV, assuring the safe handling and delivery of such food products from the point of production to the end user. Technological developments in smart packaging offer new prospects to reduce losses, maintain quality, add value and extend shelf-life of agricultural produce [105]. More novel and emerging packaging technologies are therefore still needed in the way we handle and package FFV to meet the increasing consumer demand for consistent supply of high quality, wholesome and nutritious products.

Smart, active and intelligent packaging with food spoilage indicator label (Green = fresh; orange = warning) are beginning to emerge in FFV industry. We are also beginning to see freshness and leakage indicators are commercially available for monitoring food [132]. Recent advances in biotechnology, nanotechnology, nano-sensors and material science offer new opportunity to develop new packaging materials and design for the fresh-cut fruit and vegetable industry. Incorporation of nano-sensors in the packaging material could capture and analyze environmental signals and adjust stress response treatments on fresh fruits and vegetables. Evidently, recent developments and applications of nanotechnology could lead to the development of antimicrobial packaging in response to spoilage. As stated by Opara and Mditshwa [105], the application of emerging technologies in packaging design offers new prospects for advanced quality monitoring using electronic devices that monitor and report real time information on nutritional quality and safety of food. This and other areas of packaging design remain a challenge for food, chemical and mechanical engineers.

7. Conclusion

FFV are increasing in demand due to its less processing and high nutritional content. However, the impact of processing and storage conditions should be taken into consideration by consumers due to the fact that nutritional quality of the produce can change as a result of storage and due largely to biochemical and enzymatic reactions. While conventional food processing method extends the shelf life and wholesomeness of fruits and vegetables, fresh-cut processing renders the products highly perishable and undesirable by the consumers. Suitable technology and techniques for preservation, retention of wholesomeness and consumer desirability of fresh-cut products are therefore required to meet the present day growing consumer demands.

Author details

Afam I.O. Jideani[1], Tonna A. Anyasi[1]*, Godwin R.A. Mchau[2], Elohor O. Udoro[1] and Oluwatoyin O. Onipe[1]

*Address all correspondence to: tonna.anyasi@univen.ac.za; tonna.anyasi@gmail.com

1 Department of Food Science and Technology, School of Agriculture, University of Venda, Thohoyandou, Limpopo Province, South Africa

2 Department of Horticultural Sciences, School of Agriculture, University of Venda, Thohoyandou, Limpopo Province, South Africa

References

[1] Rico D, Martin-Diana AB, Barat JM, Barry-Ryan C. Extending and measuring the quality of fresh-cut fruit and vegetables: A review. Trends in Food Science & Technology. 2007;**18**:373-386

[2] Pradas-Baena I, Moreno-Rojas JM, Luque de Castro MD. Effect of processing on active compounds in fresh-cut vegetables. In: Preedy V, editors. Processing and Impact on Active Components in Food. Oxford, UK: Academic Press, Elsevier Inc; 2015. pp. 3-10. DOI: 10.1016/B978-0-12-404699-3.00001-9

[3] Gonzalez-Aguilar GA, Ayala-Zavala JF, Ruiz-Cruz S, Acedo-Felix E, Díaz-Cinco ME. Effect of temperature and modified atmosphere packaging on overall quality of fresh-cut bell peppers. LWT. 2004;**37**(8):817-826

[4]   Smetanska I, Hunaefi D, Barbosa-Canovas GV. Nonthermal technologies to extend the shelf life of fresh-cut fruits and vegetables. In: Yanniotis S, Taoukis P, Stoforos N, Karathanos VT, editors. Advances in Food Process Engineering Research and Applications, Food Engineering Series. New York, USA: Springer Science+Business Media; 2013. pp. 375-413. DOI: 10.1007/978-1-4614-7906-2_18

[5]   Fonseca SC, Oliveira FAR, Brecht JK, Chau KV. Development of perforation-mediated atmosphere packaging for fresh-cut vegetables. In: Oliveira FAR, Oliveira JC, Hendrickx ME, Korr D, Gorris LGM, editors. Processing Foods: Quality Optimization and Process Assessment. Boca Raton: CRC Press; 1999

[6]   Artes-Hernandez P, Gomez PA, Artes F. Unit processing operations in the fresh-cut horticultural products industry: Quality and safety preservation. In: Lima GPP, Vianello F, editors. Food Quality, Safety and Technology. Vienna, Austria: Springer-Verlag; 2013. pp. 35-52. DOI: 10.1007/978-3-7091-1640-1_3

[7]   Artes F, Gomez P, Aguayo E, Escalona V, Artes-Hernandez F. Sustainable sanitation techniques for keeping quality and safety of fresh-cut plant commodities. Postharvest Biology and Technology. 2009;**51**:287-296

[8]   Sandhya M. Modified atmosphere packaging of fresh produce: Current status and future needs. LWT- Food Science and Technology. 2010;**43**(3):381-392

[9]   Toivonen PMA, Brummell DA. Biochemical bases of appearance and texture changes in fresh-cut fruit and vegetables. Postharvest Biology and Technology. 2008;**48**:1-14

[10] Kays SJ. Preharvest factors affecting appearance. Postharvest Biology and Technology. 1999;**15**:233-247

[11] Lante A, Nicoletto FTM. UV-A light treatment for controlling enzymatic browning of fresh-cut fruits. Innovative Food Science and Emerging Technologies. 2016;**34**:141-147

[12] Greenwald P, Clifford CK, Milner JA. Diet and cancer prevention. European Journal of Cancer. 2001;**37**:948-965

[13] Temple NJ, Gladwin KK. Fruit, vegetables, and the prevention of cancer: Research challenges. Nutrition. 2003;**19**:467-470

[14] Li X, Qinghong Long Q, Gao F, Han C, Jin P, Zheng Y. Effect of cutting styles on quality and antioxidant activity in fresh-cut pitaya fruit. Postharvest Biology and Technology. 2017;**124**:1-7

[15] Degl'Innocenti E, Pardossi A, Tognoni F, Guidi L. Physiological basis of sensitivity to enzymatic browning in 'lettuce', 'escarole' and 'rocket salad' when stored as fresh-cut products. Food Chemistry. 2007;**104**(1):209-215

[16] Del Nobile MA, Baiano A, Benedetto A, Massignan L. Respiration rate of minimally processed lettuce as affected by packaging. Journal of Food Engineering. 2006;**74**(1):60-69

[17] Aguayo E, Escalona V, Artes F. Metabolic behavior and quality changes of whole and fresh processed melon. Journal of Food Science. 2004;**69**:149-155

[18] Hodges DM, Toivonen PMA. Quality of fresh-cut fruits and vegetables as affected by exposure to abiotic stress. Postharvest Biology and Technology. 2008;**48**:155-162

[19] Portela SI, Cantwell MI. Cutting blade sharpness affects appearance and other quality attributes of fresh-cut cantaloupe melon. Journal of Food Science. 2001;**66**:1265-1270

[20] Bolin HR, Huxsoll CC. Control of minimally processed carrot (Daucus carota) surface discoloration caused by abrasion peeling. Journal of Food Science. 1991;**56**:416-418

[21] Tatsumi Y, Watada A, Wergin W. Scanning electron microscopy of carrot stick surface to determine the cause of white translucent appearance. Journal of Food Science. 1991;**56**:1357-1359

[22] Barry-Ryan C, O'Beirne D. Quality and shelf-life of fresh cut carrot slices as affected by slicing method. Journal of Food Science. 1998;**63**:851-856

[23] Myung K, Hamilton-Kemp TR, Archbold DD. Biosynthesis of trans-2-hexenal in response to wounding in strawberry fruit. Journal of Agricultural and Food Chemistry. 2006;**54**:1442-1448

[24] Saltveit ME. Wound induced changes in phenolic metabolism and tissue browning are altered by heat shock. Postharvest Biology and Technology. 2000;**21**(1):61-69

[25] Rakwal R, Agrawal GK. Wound signaling-coordination of the octadecanoid and mapk pathways. Plant Physiology and Biochemistry. 2003;**41**(10):855-861

[26] Cisneros-Zevallos L. The use of controlled postharvest abiotic stresses as a tool for enhancing the nutraceutical content and adding-value of fresh fruits and vegetables. Journal of Food Science. 2003;**68**:1560-1565

[27] Mishra BB, Gautam S, Sharma A. Browning of fresh-cut eggplant: Impact of cutting and storage. Postharvest Biology and Technology. 2012;**67**:44-51

[28] Alarcón-Flores MI, Romero-González R, Vidal JLM, González FJE, Frenich AG. Monitoring of phytochemicals in fresh and fresh-cut vegetables: A comparison. Food Chemistry. 2014;**142**:392-399

[29] Kenny O, O'Beirne D. Antioxidant phytochemicals in fresh-cut carrot disks as affected by peeling method. Postharvest Biology and Technology. 2010;**58**:247-253

[30] Robles-Sánchez RM, Rojas-Graüb MA, Odriozola-Serrano I, González-Aguilar GA, Martín-Belloso O. Effect of minimal processing on bioactive compounds and antioxidant activity of fresh-cut 'Kent' mango (*Mangifera indica* L.). Postharvest Biology and Technology. 2009;**51**:384-390

[31] Son SM, Moon KD, Lee CY. Inhibitory effects of various antibrowning agents on apple slices. Food Chemistry. 2001;**73**:23-30

[32] Torres-Contreras AM, Nair V, Cisneros-Zevallos L, Jacobo-Velazquez DA. Plants as biofactories: Stress-induced production of chlorogenic acid isomers in potato tubers as affected by wounding intensity and storage time. Industrial Crops Products. 2014;**62**:61-66

[33] Surjadinata BB, Cisneros-Zevallos L. Biosynthesis of phenolic antioxidants in carrot tissue increases with wounding intensity. Food Chemistry. 2012;**134** (2):615-624

[34] Vina SZ, Chaves AR. Antioxidant responses in minimally processed celery during refrigerated storage. Food Chemistry. 2006;**94** (1):68-74

[35] Zhan LJ, Li Y, Hu JQ, Pang LY, Fan HP. Browning inhibition and quality preservation of fresh-cut romaine lettuce exposed to high intensity light. Innovative Food Science and Emerging Technology. 2012;**14**:70-76

[36] Benito Martinez-Hernandez G, Artes-Hernandez F, Gomez PA, Formica AC, Artes, F. Combination of electrolysed water: UV-C and super atmospheric O_2 packaging for improving fresh-cut broccoli quality. Postharvest Biology and Technology. 2013;**76**:125-134

[37] Maribel Robles-Sanchez R, Alejandra Rojas-Gratue M, Odriozola-Serrano I, Gonzalez-Aguilar G, Martin-Belloso O. Influence of alginate-based edible coating as carrier of antibrowning agents on bioactive compounds and antioxidant activity in fresh-cut kent mangoes. LWT—Food Science and Technology. 2013;**50**(1):240-246

[38] Kader AA. Quality parameters of fresh-cut fruit and vegetable products. In: Lamikanra O, editor. Fresh-Cut Fruits and Vegetables: Science, Technology and Safety. Boca Raton, USA: CRC Press LLC; 2002. DOI: 10.1201/9781420031874

[39] Meireles A, Giaouris E, Simões M. Alternative disinfection methods to chlorine for use in the fresh-cut industry. Food Research International. 2016;**82**:71-85

[40] Ramos B, Miller FA, Brandão TRS, Teixeira P, Silva CLM. Fresh fruits and vegetables — An overview on applied methodologies to improve its quality and safety. Innovative Food Science & Emerging Technologies. 2013;**20**:1-15

[41] Yaron S, Romling U. Biofilm formation by enteric pathogens and its role in plant colonization and persistence. Microbial Biotechnology. 2014;**7**(6):496-516

[42] Parish ME, Beuchat LR, Suslow TV, Harris LJ, Garrett EH, Farber JN, Busta FF. Methods to reduce/eliminate pathogens from fresh and fresh-cut produce. Comprehensive Reviews in Food Science and Food Safety. 2003;**2**:161-173

[43] Bull RJ, Reckhow DA, Li X, Humpage AR, Joll C, Hrudey SE. Potential carcinogenic hazards of non-regulated disinfection by-products: Haloquinones, halo-cyclopentene and cyclohexene derivatives, N-halamines, halonitriles, and heterocyclic amines. Toxicology. 2011;**286**(1-3):1-19

[44] Legay C, Rodriguez MJ, Sérodes JB, Levallois P. Estimation of chlorination by-products presence in drinking water in epidemiological studies on adverse reproductive outcomes: A review. Science of the Total Environment. 2010;**408**(3):456-472

[45] Bilek SE, Turantaş F. Decontamination efficiency of high power ultrasound in the fruit and vegetable industry, a review. International Journal of Food Microbiology. 2013;**166**(1):155-162

[46] Fallik E. Microbial quality and safety of fresh produce. In: Florkowski WJ, Shewfelt RL, Brueckner B, Prussia SE, editors. Postharvest Handling. 3rd ed. San Diego: Academic Press; 2014. pp. 313-339

[47] Ölmez H, Kretzschmar U. Potential alternative disinfection methods for organic fresh-cut industry for minimizing water consumption and environmental impact. LWT — Food Science and Technology. 2009;**42**(3):686-693

[48] Gil MI, Selma MV, López-Gálvez F, Allende A. Fresh-cut product sanitation and wash water disinfection: Problems and solutions. International Journal of Food Microbiology. 2009;**134**(1-2):37-45

[49] Goodburn C, Wallace CA. The microbiological efficacy of decontamination methodologies for fresh produce: A review. Food Control. 2013;**32**(2):418-427

[50] Holah JT. Cleaning and disinfection practices in food processing.. In: Lelieveld HLM, Holah JT, Napper D, editors. Hygiene in Food Processing. 2nd ed. Cambridge CB22 3HJ, UK: Woodhead Publishing; 2014. pp. 259-304. DOI: 10.1533/9780857098634.3.259

[51] Otto C, Zahn S, Rost F, Zahn P, Jaros D, Rohm H. Physical methods for cleaning and disinfection of surfaces. Food Engineering Reviews. 2011;**3**(3-4):171-188

[52] Lado BH, Yousef AE. Alternative food-preservation technologies: Efficacy and mechanisms. Microbes and Infection. 2002;**4**(4):433-440

[53] Tirpanalan O, Zunabovic M, Domig K, Kneifel W. Mini review: Antimicrobial strategies in the production of fresh-cut lettuce products. In: Méndez-Vilas A, editors. Science Against Microbial Pathogens: Communicating Current Research and Technological Advances 1. Badajoz, Spain: Formatex Research Center; 2011. pp. 176-188

[54] Garcia E, Barrett DM. Preservative treatments for fresh-cut fruits and vegetables. In: Lamikanra, O, editors. Fresh-cut Fruits and Vegetables Science Technology and Market. Boca Raton: CRC Press; 2002. pp. 267-304

[55] Gonzalez-Aguilar GA, Kader AA, Brecht JK, Toivonen PMA. Fresh-cut tropical and subtropical fruit products. In: Yahia EM, editors. Postharvest Biology and Technology of Tropical and Subtropical Fruits Volume 1. Cambridge, UK: Woodhead Publishing; 2011. pp. 381-418

[56] Djioua T, Charles F, Lopez-Lauri F, Filgueiras H, Coudret A, Freire Jr M, Ducamp-Collin MN, Sallanon H. Improving the storage of minimally processed mangoes (*Mangifera indica* L.) by hot water treatments. Postharvest Biology and Technology. 2009;**52**:221-226

[57] Techavuthiporn C, Boonyaritthonghai P. Effects of anti-browning agents on wound responses of fresh-cut mangoes. International Food Research Journal. 2016;**23**(5):1879-1886

[58] Plotto A, Narciso JA, Rattanapanone N, Baldwin EA. Surface treatments and coatings to maintain fresh-cut mango quality in storage. Journal of the Science Food and Agriculture. 2010;**90**:2333-2341

[59] Chiumarelli M, Ferrari CC, Sarantopoulos CIGL, Hubinger MD. Fresh cut 'Tommy Atkins' mango pre-treated with citric acid and coated with cassava (Manihot esculenta Crantz) starch or sodium alginate. Innovative Food Science and Emerging Technologies. 2011;**12**:381-387

[60] González-Aguilar GA, Wang CY, Buta JG. Maintaining quality of fresh-cut mangoes using antibrowning agents and modified atmosphere packaging. Journal of Agricultural and Food Chemistry. 2000;**48**(9):4204-4208

[61] Francis GA, Gallone A, Nychas GJ, Sofos JN, Colelli G, Amodio ML, Spano G. Factors affecting quality and safety of fresh-cut produce. Critical reviews in food science and nutrition. 2012;**52**:595-610. DOI: 10.1080/10408398.2010.503685

[62] Keith W, Huber A, Namvar A, Fan W, Dunfield K. Recent advances in the microbial safety of fresh fruits and vegetables. Advances in Food and Nutrition Research. 2009;**57**:155-208

[63] Ukuku DO, Fett W. Behaviour of Listeria monocytogenes inoculated on cantaloupe surfaces and efficacy of washing treatments to reduce transfer from rind to fresh-cut pieces. Journal of Food Protection. 2002;**65**:924-930. DOI: 10.4315/0362-028X-65.6.924

[64] Conway WS, Leverentz B, Saftner RA, Janisiewicz WJ, Sams CE, Leblanc E. Survival and growth of Listeria monocytogenes on fresh-cut apple slices and its interaction with Glomerella cingulata and Penicillium expansum. Plant Disease. 2000;**84**:177-181

[65] Liao CH, Sapers GM. Attachment and growth of Salmonella Chester on apple fruits and in vivo response of attached bacteria to sanitizer treatments. Journal of Food Protection. 2000;**63**:876-883. http://dx.doi.org/10.4315/0362-028X-63.7.876

[66] Francis GA, O'Beirne D. Effects of the indigenous microflora of minimally processed lettuce on the survival and growth of *Listeria innocua*. International Journal of Food Science & Technology. 1998;**33**:477-488. DOI: 10.1046/j.1365-2621.1998.00199.x

[67] Gracias KS, McKillip JL. A review of conventional detection and enumeration methods for pathogenic bacteria in food. Canadian Journal of Microbiology. 2004;**50**:883-890

[68] Manafi M. Fluorogenic and chromogenic enzyme substrates in culture media and identification tests. International Journal of Food Microbiology. 1996;**31**:45-58. http://dx.doi.org/10.1016/0168-1605(96)00963-4

[69] ISO. Microbiology of Food and Animal Feeding Stuffs- Horizontal Method for the Detection and Enumeration of Listeria Monocytogenes. Part 2: Enumeration Method. International Standard ISO 11290-2. ed. Geneva. Switzerland: International Organization for Standardization; 1996

[70] EC. Opinion of the Scientific Committee on Veterinary Measures relating to Public Health on Listeria Monocytogenes. Brussels, Belgium: European Commission, Health and Consumer Protection Directorate-General; 1999

[71] Ponniah J, Robin T, Paie MS, Radu S, Mohamad Ghazali F, Cheah YK. Detection of Listeria monocytogenes in foods. International Food Research Journal. 2010;**17**:1-1

[72] Foley SL, Grant K. Molecular techniques of detection and discrimination of foodborne pathogens and their toxins. In: Simjee S, editors. Foodborne Diseases. Totowa, NJ: Humana Press; 2007. pp. 485-510

[73] Harris L, Farber J, Beuchat L, Parish M, Suslow T, Garrett E, Busta F. Outbreaks associated with fresh produce: Incidence, growth, and survival of pathogens in fresh and fresh-cut produce. Comprehensive Reviews in Food Science and Food Safety. 2003;**2**(s1):78-141

[74] Velderrain-Rodríguez GR, Quirós-Sauceda AE, González Aguilar GA, Siddiqui MW, Zavala JFA. Technologies in fresh-cut fruit and vegetables. In: Siddiqui MW, Rahman MS, editors. Minimally Processed Foods, Food Engineering Series. Switzerland: Springer International Publishing; 2015. DOI: 10.1007/978-3-319-10677-9_5

[75] Saranraj P. Microbial spoilage of bakery products and its control by preservatives. International Journal of Pharmaceutical and Biological Archive. 2012;**3**(1):38-48

[76] Olaimat AN, Holley RA. Factors influencing the microbial safety of fresh produce: A review. Food Microbiology. 2012;**32**(1):1-19

[77] Wirtanen G, Salo S, Helander I, Mattila-Sandholm T. Microbiological methods for testing disinfectant efficiency on Pseudomonas biofilm. Colloids and Surfaces B: Biointerfaces. 2001;**20**(1):37-50

[78] James G, Beaudette L, Costerton J. Interspecies bacterial interactions in biofilms. Journal of Industrial Microbiology. 1995;**15**(4):257-262

[79] Tomas-Callejas A, Lopez-Velasco G, Sbodio A, Artes F, Artes-Hernandez F, Suslow TV. Survival and distribution of *Escherichia coli* on diverse fresh-cut baby leafy greens under preharvest through postharvest conditions. International Journal of Food Microbiology. 2011;**151**:216-222

[80] Vitale M, Schillaci D. Food Processing and Foodborne Illness. Reference Module in Food Science. Elsevier; 2016

[81] Golberg D, Kroupitski Y, Belausov E, Pinto R, Sela S. Salmonella Typhimurium internalization is variable in leafy vegetables and fresh herbs. International Journal of Food Microbiology. 2011;**145**(1):250-257

[82] Seo KH, Frank JF. Attachment of *Escherichia coli* O157:H7 to lettuce leaf surface and bacterial viability in response to chlorine treatment as demonstrated by using confocal scanning laser microscopy. Journal of Food Protection. 1999;**62**(1):3-9. http://jfoodprotection.org/doi/abs/10.4315/0362-028X-62.1.3?code=fopr-site

[83] Erickson MC. Internalization of fresh produce by foodborne pathogens. Annual Review of Food Science and Technology. 2012;3:283-310

[84] Ge C, Bohrerova Z, Lee J. Inactivation of internalized Salmonella Typhimurium in lettuce and green onion using ultraviolet C irradiation and chemical sanitizers. Journal of Applied Microbiology. 2013;114(5):1415-1424

[85] Rodgers S. Novel applications of live bacteria in food services: Probiotics and protective cultures. Trends in Food Science & Technology. 2008;19(4):188-197

[86] Arevalos-Sánchez M, Regalado C, Martin SE, Domínguez-Domínguez J, García-Almendárez BE. Effect of neutral electrolyzed water and nisin on Listeria monocytogenes biofilms, and on listeriolysin O activity. Food Control. 2012;24(1-2):116-122

[87] Magalhães R, Mena C, Ferreira V, Silva J, Almeida G, Gibbs P, Teixeira P. Bacteria: Listeria monocytogenes. In: Motarjemi Y, editor. Encyclopedia of Food Safety. Waltham: Academic Press; 2014. pp. 450-461

[88] Bari ML, Ukuku DO, Kawasaki T, Inatsu Y, Isshiki K, Kawamoto S. Combined efficacy of nisin and pediocin with sodium lactate, citric acid, phytic acid, and potassium sorbate and EDTA in reducing the Listeria monocytogenes population of inoculated fresh-cut produce. Journal of Food Protection. 2005;68(7):1381-1387

[89] Spricigo DA, Bardina C, Cortés P, Llagostera M. Use of a bacteriophage cocktail to control Salmonella in food and the food industry. International Journal of Food Microbiology. 2013;165(2):169-174

[90] Sharma M, Ryu JH, Beuchat LR. Inactivation of *Escherichia coli* O157:H7 in biofilm on stainless steel by treatment with an alkaline cleaner and a bacteriophage. Journal of Applied Microbiology. 2005;99(3):449-459

[91] Simões M, Simões LC, Vieira MJ. A review of current and emergent biofilm control strategies. LWT — Food Science and Technology. 2010;43(4):573-583

[92] Thallinger B, Prasetyo EN, Nyanhongo GS, Guebitz GM. Antimicrobial enzymes: An emerging strategy to fight microbes and microbial biofilms. Biotechnology Journal. 2013;8(1):97-109

[93] Lequette Y, Boels G, Clarisse M, Faille C. Using enzymes to remove biofilms of bacterial isolates sampled in the food-industry. Biofouling. 2010;26(4):421-431

[94] Augustin M, Ali-Vehmas T, Atroshi F. Assessment of enzymatic cleaning agents and disinfectants against bacterial biofilms. Journal of Pharmacy & Pharmaceutical Sciences. 2004;7(1):55-64

[95] Farber, JN, Harris, LJ, Parish, ME, Beuchat, LR, Suslow, TV, Gorney, JR, Garrett, EH, Busta, FF. Microbiological safety of controlled and modified atmosphere packaging of fresh and fresh-cut produce. Comprehensive Review in Food Science and Food Safety. 2003;2:142-160

[96] Mahajan PV, Oliveira FAR, Montanez JC, Frias J. Development of user-friendly soft-ware for design of modified atmosphere packaging for fresh and fresh-cut produce. Innovative Food Science and Emerging Technologies. 2007; 8:84-92

[97] Caleb OJ, Opara UL, Witthuhn CR. Modified atmosphere packaging of pomegranate fruit and arils: A review. Food and Bioprocess Technology. 2012;5:15-30. DOI: 10.1007/s11947-011-0525-7

[98] Caleb OJ, Mahajan PV, Al-Said F.Al-J, Opara UL. Modified atmosphere packaging technology of fresh and fresh-cut produce and the microbial consequences—A review. Food and Bioprocess Technology. 2013;6:303-329. DOI: 10.1007/s11947-012-0932-4

[99] Kader, AA, Watkins, CB. Modified atmosphere packaging-toward 2000 and beyond. HortTechnology. 2000;10(3):483-486

[100] Charles F, Sanchez J, Gontard N. Active modified atmosphere packaging of fresh fruits and vegetables: Modeling with tomatoes and oxygen absorber. Journal of Food Science. 2003;68(5):1736-1742

[101] Gorny JR, Hess-Pierce B, Cifuentes RA, Kader AA. Quality changes in fresh-cut pear slices as affected by controlled atmospheres and chemical preservatives. Postharvest Biology and Technology. 2002;24(3):271-278

[102] Soliva-Fortuny RC, Elez-Martínez P, Martín-Belloso O. Microbiological and bio-chemical stability of fresh-cut apples preserved by modified atmosphere packaging. Innovative Food Science & Emerging Technologies. 2004;5(2):215-224

[103] Gorny JR, Hess-Pierce B, Kader AA. Quality changes in fresh-cut peach and nectarine slices as affected by cultivar, storage atmosphere and chemical treatments. Journal of Food Science. 1999;64(3):429-432

[104] Maghenzani M, Chiabrando V, Giacalone G. The effect of different MAP on quality reten-tion of fresh-cut nectarines. International Food Research Journal. 2016;23(5):1872-1878

[105] Opara UL, Mditshwa A. A review on the role of packaging in securing food system: Adding value to food products and reducing losses and waste. African Journal of Agricultural Research. 2013;8(22):2621-2630. DOI: 10.5897/AJAR2013.6931

[106] Ares, G, De Saldamando L, Giménez A, Deliza R. Food and wellbeing: Towards a con-sumer-based approach. Appetite. 2014;74:61-69

[107] Haynes-Maslow L, Parsons SE, Wheeler SB, Leone LA. A qualitative study of perceived barriers to fruit and vegetable consumption among low-income populations, North Carolina, 2011. Preventing Chronic Disease. 2013;10:1-10

[108] Li L, Pegg RB, Eitenmiller RR, Chun Ji-Y, Kerrihard AL. Selected nutrient analyses of fresh, fresh-stored, and frozen fruits and vegetables. Journal of Food Composition and Analysis. 2017;59:8-17

[109] De Anos B, Sanchez-Moreno C, Pascual-Teresa S, Cano MP. Freezing preservation of fruits. In: Sinha N, Sidhu JS, Barta J, Wu JSB, Cano MP, editors. Handbook of Fruits and Fruits Processing. 2ndnd ed. Oxford, UK: John Wiley & Sons; 2012. pp. 103-119

[110] Jaiswal AK, Gupta S, Abu-Ghannam N. Kinetic evaluation of colour, texture, polyphenols and antioxidant capacity of Irish York cabbage after blanching treatment. Food Chemistry. 2012;**131**:63-72

[111] Van Buggenhout S, Lille M, Messagie I, Van Loey A, Autio K, Hendrickx M. Impact of pretreatment and freezing conditions on the microstructure of frozen carrots: Quantification and relation to texture loss. European Food Research and Technology. 2006;**222**:543-553

[112] Paciulli M, Ganino T, Pellegrini N, Rinaldi M, Zaupa M, Fabbri A, Chiavaro E. Impact of the industrial freezing process on selected vegetables — Part I. Structure, texture and antioxidant capacity. Food Research International. 2015;**74**:329-337

[113] O'Beirne D, Francis GA. Reducing pathogen risk in MAP-prepared produce. In: Ahvenainen R, editor. Novel Food Packaging Techniques. Boca Raton, FL: Woodhead Publishing Limited/CRC Press LLC; 2003. pp. 231-232

[114] Dermesonlouoglou EK, Giannakourou MC, Taoukis PS. Kinetic modeling of the degradation of quality of osmo-dehydrofrozen tomatoes during storage. Food Chemistry. 2007;**103**:985-993

[115] Said LBH, Bellagha S, Allaf K. Dehydrofreezing of apple fruits: Freezing profiles, freezing characteristics, and texture variation. Food and Bioprocess Technology. 2016;**9**:252-261

[116] Waldron KW, Parker ML, Smith AC. Plant cell walls and food quality. Comprehensive Reviews in Food Science and Food Safety. 2003;**2**:128-146

[117] James C, Purnel G, James SJ. A critical review of dehydrofreezing of fruits and vegetables. Food and Bioprocess Technology. 2014;**7**(5)DOI: 10.1007/s11947-014-1293-y

[118] Li B, Sun D.-W. Novel methods for rapid freezing and thawing of foods—A review. Journal of Food Engineering. 2002;**54**(3):175-182. DOI: 10.1016/S0260-8774(01)00209-6

[119] Wu L, Orikasa T, Tokuyasu K, Shiina T, Tagawa A. Applicability of vacuum-dehydrofreezing technique for the long term preservation of fresh-cut eggplant: Effects of process conditions on the quality attributes of the samples. Journal of Food Engineering. 2009;**91**(4):560-565. DOI: 10.1016/j.jfoodeng.2008.10.021

[120] Ramallo LA, Mascheroni RH. Dehydrofreezing of pineapple. Journal of Food Engineering. 2010;**99**(3):269-275. DOI: 10.1016/j.jfoodeng.2010.02.026

[121] Barrett DM, Beaulieu JC, Shewfelt R. Color, flavor, texture, and nutritional quality of fresh-cut fruits and vegetables: Desirable levels, instrumental and sensory measurement, and the effects of processing. Critical Reviews in Food Science and Nutrition. 2010;**50**:369-389. DOI: 10.1080/10408391003626322

[122] Benítez S, Achaerandio I, Pujolà M, Sepulcre F. Aloe vera as an alternative to traditional edible coatings used in fresh-cut fruits: A case of study with kiwifruit slices. LWT-Food Science and Technology. 2015;61(184-193). http://dx.doi.org/10.1016/j.lwt.2014.11.036

[123] Li X, Long Q, Gao F, Han C, Jin P, Zheng Y. Effect of cutting styles on quality and antioxidant activity in fresh-cut pitaya fruit. Postharvest Biology and Technology. 2017;124(1-7). http://dx.doi.org/10.1016/j.postharvbio.2016.09.009

[124] Helland HS, Leufvén A, Bengtsson GB, Pettersen MK, Lea P, Wold AB. Storage of fresh-cut swede and turnip: Effect of temperature, including sub-zero temperature, and packaging material on sensory attributes, sugars and glucosinolates. Postharvest Biology and Technology. 2016;111:370-379. http://dx.doi.org/10.1016/j.postharvbio.2015.09.011

[125] Kim MJ, Bang WS, Yuk HG. 405 ± 5 nm light emitting diode illumination causes photodynamic inactivation of Salmonella spp. on fresh-cut papaya without deterioration. Food Microbiology. 2017;62:124-132. http://dx.doi.org/10.1016/j.fm.2016.10.002

[126] Chen C, Hu W, He Y, Jiang A, Zhang R. Effect of citric acid combined with UV-C on the quality of fresh-cut apples. Postharvest Biology and Technology. 2016;111:126-131. http://dx.doi.org/10.1016/j.postharvbio.2015.08.005

[127] Moreira MR, Tomadoni B, Martín-Belloso O, Soliva-Fortuny R. Preservation of fresh-cut apple quality attributes by pulsed light in combination with gellan gum-based prebiotic edible coatings. LWT – Food Science and Technology. 2015;65:1130-1137. http://dx.doi.org/10.1016/j.lwt.2015.07.002

[128] Kadzere I, Watkins CB, Merwin IA, Akinnifesi FK, Saka JDK, Mhango J. Harvesting and postharvest handling practices and characteristics of *Uapaca kirkiana* (*Muell. arg.*) fruits: a survey of roadside markets in Malawi. Agroforest System. 2006;68:133-142. DOI: 10.1007/s10457-006-9004-y.

[129] Masarirambi MT, Mavuso V, Songwe VD, Nkambule TP, Mhazo N. Indigenous post-harvest handling and processing of traditional vegetables in Swaziland: A review. African Journal of Agricultural Research. 2010;5(24):3333-3341. DOI: 10.5897/AJAR10.685

[130] Nur Arina AJ, Azrina A. Comparison of phenolic content and antioxidant activity of fresh and fried local fruits. International Food Research Journal. 2016;23(4):1717-1724

[131] National Institute of Health. Antioxidant [Internet]. 2012. Available from: http://www.nlm.nih.gov/medlineplus/antioxidants.html [Accessed: 14-May-2017]

[132] Nopwinyuwong A, Trevanich S, Suppakul P. Development of a novel colorimetric indicator label for monitoring freshness of intermediate-moisture dessert spoilage. Talanta. 2010;81(3):1126-1132

7

Fresh-Cut Fruit and Vegetables: Emerging Eco-friendly Techniques for Sanitation and Preserving Safety

Francisco Artés-Hernández,
Ginés Benito Martínez-Hernández, Encarna Aguayo,
Perla A. Gómez and Francisco Artés

Additional information is available at the end of the chapter

Abstract

The current high demand of minimally processed or fresh-cut fruit and vegetables results from the consumer's desire for healthy, convenient, fresh, and ready-to-eat plant food-derived commodities. Fresh-cut fruits and vegetables are usually packaged under active- or passive-modified atmosphere packaging, while its shelf life must be under refrigerated conditions. The most important goal to preserve quality and safety focuses on releasing the microbial spoilage flora, since every unit operation involved will influence the final load. Sanitation in the washing step is the only unit operation able to reduce microbial load throughout the production chain. Chlorine is widely used as an efficient sanitation agent, but some disadvantages force to find eco-friendly emerging alternatives. It is necessary to deal with aspects related to sustainability because it could positively contribute to the net carbon balance besides reducing its use. Several innovative techniques seem to reach that target. However, industrial changes for replacing conventional techniques request a fine knowledge of the benefits and restrictions as well as a practical outlook. This chapter reviews the principles of emerging eco-friendly techniques for preserving quality and safety of fresh-cut products in order to meet the expected market's demand.

Keywords: minimally processed, ready-to-eat, sanitizing, pathogens, food safety

1. Introduction

The benefits of fruit and vegetables consumption on human health are well known, being linked with prevention of a grand array of diseases such as degenerative disorders, cancer, and cardiovascular, among others [1]. Consequently, their intake has been promoted among consumers by nutritionists, researchers, and even at a governmental level (i.e., campaigns like *five-a-day*, etc.). However, the actual consumer's demand of new food was elaborated by the industry with the following characteristics: freshness, healthiness, and easy- or ready-to-eat. Particularly, minimally processed or fresh-cut (FC) fruit and vegetables connect well within such consumer needs. The main advantage of FC plant foods is that they have nearly the same properties as the whole intact product, but they do not need much elaboration time and are with a uniform and consistent quality [2]. NaOCl has been widely used in the FC industry as a strong sanitizing agent due to its powerful oxidizing properties [3]. Among the main NaOCl advantages are high effectiveness, comparatively inexpensive, and that they may be implemented in any size operations [4]. Nevertheless, NaOCl may produce unhealthy by-products in processed water (chloramines, chloroform, haloacetic acids, or other trihalo-methanes) that have been reported to present carcinogenic or mutagenic effects, with proven toxicity to liver and kidneys [3]. Therefore, NaOCl use in the FC industry has been forbidden in some European countries [5].

This chapter summarizes the principles and development of eco-friendly techniques for preserving safety of FC products in order to meet the expected market's demand.

2. Antimicrobial washing solutions

2.1. Peroxyacetic acid

Peroxyacetic acid (PAA), or peracetic acid, is a colorless organic peroxide that is a mixture of acetic acid and hydrogen peroxide. It is an eco-friendly sanitizer whose breakdown products are acetic acid, O_2, CO_2, and water. PAA is approved by the European Union as a disinfectant for drinking water and food areas and as an in-can preservative [6]. PAA is also permitted by the U.S. Food and Drug Administration as an additive for food [7]. The surface-cleaning concentrations range from 85 to 300 ppm, although 50 ppm has been reported to be enough [8]. PAA is mainly used in fruit and vegetable processing due to tolerance to several factors such as temperature, pH (1–8), hardness, and soil contamination. A recommended combination of 11% hydrogen peroxide (H_2O_2) and 15% PAA, at 80 ppm, was proposed for the disinfection of plant surfaces [4]. *Escherichia coli* O157:H7 and *Salmonella* spp. reductions of 2–3 log CFU g^{-1} were reported in apples and melons treated with 70 ppm of PAA [9, 10]. Similarly, *E. coli* O157:H7, *Salmonella* spp., and *Listeria monocytogenes* inoculated on FC carrot shreds were reduced after PAA washing at 40 ppm for 2 min [11]. Mesophilic and psychrotrophic loads of FC Galia melon were reduced by 1 and 2 log CFU g^{-1}, respectively, using PAA (68 ppm) [12]. The nutritional and sensory quality of FC iceberg lettuce was not affected by PAA (120 ppm), while natural microflora was reduced by approximately 1 log CFU g^{-1} [13]. A similar *Salmonella typhimurium* reduction was achieved after the PAA treatment (40 ppm) in inoculated lettuce [14]. A PAA treatment of 80 ppm was more effective than 106 ppm of NaOCl to reduce *E. coli* O157:H7 and *Salmonella enterica* Montevideo on mung bean sprouts [15]. *E. coli* and *Salmonella enteritidis* reductions of 2–3 log CFU g^{-1} were achieved in the kailan-hybrid broccoli with 100 ppm of PAA being more effective than 100 ppm of NaOCl [16].

2.2. Chlorine dioxide

Chlorine dioxide (ClO_2) is a yellowish-green stable dissolved gas that has been used for the last decades for water treatment as a potential alternative to NaOCl. ClO_2 has higher effectiveness over a broad range of pH, higher water solubility (10 times higher than NaOCl), higher oxidant capacity, lower reactivity with organic matter, and higher effectiveness at low concentrations than NaOCl [17]. Nevertheless, ClO_2 is a very unstable substance and is highly explosive as a concentrated gas when concentrations ≥10% are reached in air. Hence, ClO_2 must be generated on-site by two different procedures: reacting an acid with sodium chlorite or the reaction of sodium chlorite with chlorine gas then being obtained in either aqueous or gaseous forms, respectively [18]. The ClO_2 is classified as a non-carcinogenic product since it does not ionize to produce weak acids (as occurred for chlorine and bromine) or to form carcinogenic by-products like trihalomethanes [19]. Gaseous ClO_2 treatment (100 ppm) of several fresh products (tomatoes, lettuce, cantaloupe, alfalfa sprouts, oranges, apples, and strawberries) did not leave any chemical residues on them [20]. ClO_2 is approved in the USA for usage in washing whole fresh fruits and vegetables and shelled beans and pears with intact cuticles at maximum levels of 5 ppm and 1 ppm for peeled potatoes [19].

ClO_2 is considered as a strong microbicide at low levels such as 0.1 ppm, achieving also a rapid removal of biofilms which avoid bacterial re-growth [21]. The bactericidal effect of ClO_2 is explained by the interruption of several cellular processes (proteins production and changes in the cell structure) when organic substances in bacterial cells react with ClO_2. On viruses, ClO_2 reacts with peptone to prevent the protein formation being more effective than chlorine or ozone [21]. Inoculated pathogens like *Salmonella* spp., *E. coli* O157:H7, and *L. monocytogenes* were reduced on cabbage, carrot, lettuce, strawberry, and melon with ClO_2 concentrations of 4–5 ppm [22–26]. ClO_2 treatment at 100 ppm of FC cucumber, lettuce, carrot, apple, tomato, and guava reduced total bacterial and coliform counts up to 3.5–4.0 log CFU g^{-1} being more effective than the same NaOCl concentration [27]. A ClO_2 treatment of 3 ppm substantially prevented *E. coli* O157:H7 cross-contamination but was not effective for the inoculated *Salmonella* in FC Red Chard [28]. The effectiveness of ClO_2 treatment of tomato processing water under a range of water quality and temperature was studied, which showed that an increase in temperature and ClO_2 concentration reduced the contact time achieving a 6-log reduction of *S. enterica* within 2 min of contact time [29]. Acidified sodium chlorite (100–500 ppm) at low-moderate doses showed an initial antimicrobial efficacy on natural microflora and *E. coli* of FC tatsoi baby leaves as effective as that of 100 ppm NaOCl [30].

2.3. Hydrogen peroxide

Hydrogen peroxide (H_2O_2) is a strong oxidizer able to generate other cytotoxic-oxidizing species, like hydroxyl radicals, with strong bactericide (including spores) effect [31]. H_2O_2 is an eco-friendly disinfectant since it is rapidly decomposed into water and oxygen in the presence of catalase. Likewise, it is colorless and non-corrosive. H_2O_2 is allowed for use in food processing and packaging but not as a sanitizing agent for fresh produce by the FDA [5]. However, high H_2O_2 concentrations are needed to achieve good santising effects in FC products. However, such high concentrations may lead to browning being necessary thge use of anti-browning agents like sodium erythorbate [32]. Accordingly, $2\text{–}3 \times 10^4$ ppm H_2O_2 were needed to reduce *E. coli* O157:H7 by 1.6 log CFU g^{-1} in baby spinach [33]. *L. monocytogenes* reductions of 2.0–3.5 log CFU cm^{-2} were reported in melon surfaces after 5×10^4 ppm H_2O_2 treatment [34]. Effectiveness of H_2O_2 treatment (3%) on inoculated *E. coli* O157:H7, *Salmonella*, and *L. monocytogenes* in whole cantaloupe rind surfaces was enhanced when applied at 80°C for 300 s [35]. H_2O_2 has been also found to extend the shelf life and reducing native microbial and pathogen populations in whole grapes, prunes, apples, oranges, mushrooms, melons, tomatoes, red bell peppers and lettuce, and in FC cucumber, zucchini, bell peppers, and melons [5, 36]. Nevertheless, the cross-contamination may not be avoided with H_2O_2, since it may still occur in the product washing water and its breakdown is rapid with low disinfection kinetics [37].

2.4. Weak organic acids

Weak organic acids have been widely used as preservatives for the prevention of several quality degradation processes such as enzymatic and nonenzymatic browning, texture deterioration, and microbial spoilage. Contrary to NaOCl, weak organic acids do not produce toxic or carcinogenic compounds when they interact with organic molecules [38]. Therefore, several weak organic acids are considered as GRAS (Generally Recognized as Safe) by the FDA and European Commission being well accepted by consumers. The antimicrobial effect of weak organic acids is related to the cytoplasm acidification, osmotic stress, disruption of proton motive force, and synthesis inhibition of macromolecules [39]. Weak organic acids are more effective for bacteria than for yeasts and molds because of the low pH (2.1–2.7) of the applied solutions. Citric, acetic, lactic, and ascorbic acids are the most common acids applied in the food industry.

Citric acid, contrary to other acids, acts as a chelating agent of metallic ions of the medium, avoiding microbial growth [40, 41]. Citric acid treatment (0.52 mM) maintained microbial safety and visual quality of FC "Amarillo" melon during a shelf life of 10 days at 5°C [42]. A solution of 0.1 M citric and 0.5 M ascorbic acid achieved the same effectivity as 100 ppm NaOCl to control microbial growth and maintain quality of green celery crescents [43]. Citric and lactic acid dippings of $0.5\text{–}1 \times 10^4$ ppm achieved comparable *E. coli* reductions of 1.9–2.3 log CFU g^{-1} to 100 ppm NaOCl in inoculated FC lettuce without significant efficacy enhancement from incrementing dipping times from 2 to 5 min [44]. Likewise, acetic and citric acid dippings of $0.5\text{–}1 \times 10^4$ ppm achieved similar *L. monocytogenes* reductions of 0.8–1.0 log CFU g^{-1} to 100 ppm NaOCl in inoculated FC lettuce [44]. However, acetic acid and ascorbic acid dippings of $0.5\text{–}1 \times 10^4$ ppm achieved lower *E. coli* reductions than 100 ppm NaOCl in inoculated FC lettuce [44]. The effectiveness of citric, acetic, lactic, malic, and propionic acid dippings (1×10^4 ppm) for inoculated *E. coli* O157:H7, *L. monocytogenes*, and *S. typhimurium* onto fresh lettuce was studied with reductions of 1.9–2.9, 1.1–1.7, 1.9–2.5, 2.3–3.0, and 0.9–1.5 log CFU g^{-1}, respectively [45].

2.5. Calcium, sodium, and potassium-derived salts

Several salts are recognized as GRAS being a low-cost material for the food industry with high acceptance by the consumers, since it is not toxic. Calcium is used to retain the firmness of plant commodities by interaction with pectin to form calcium pectate maintaining then the cell wall structure. FC lettuce treated with 15×10^3 ppm calcium lactate showed higher crispness than samples treated with 120 ppm NaOCl after 1 day at 4°C [46]. Latter authors hypothesized that such finding could be owed to the activation of texture-related enzymes, like PME, by the calcium absorption in the lettuce, or an increase in diffusive processes by temperature, including the calcium. Similar results were obtained by 15×10^3 ppm calcium lactate treatment at 50°C of sliced FC carrots to maintain the cortex turgor of plant cells and reduce the lignification degree in cut surfaces [47]. However, the calcium lactate antimicrobial properties have been scarcely studied. FC lettuce and carrots treated with 3×10^4 ppm calcium lactate showed similar microbial loads than 120 ppm NaOCl after 10 days at 4°C [48]. Similarly, 3×10^4 ppm calcium lactate treatment at 60°C of FC melon induced 1–2 log CFU g^{-1} lower bacterial and yeasts and mold loads after 8 days at 5°C, while texture of such melon pieces was better maintained than samples washed with water at the same temperature [49]. Latter authors also found that other calcium salt treatments (calcium chloride and calcium propionate) maintained lower microbial loads in the FC melon samples after 8 days at 5°C. Calcium pre-treatments may also help to prevent enzymatic browning reactions during high-pressure processing (HPP) of peaches if the penetration of Ca^{2+} reaches the target area, which is the tonoplast or vacuolar membrane as observed in peaches [50].

Sodium bicarbonate, sodium carbonate, sodium silicate, potassium bicarbonate, potassium carbonate, and potassium sorbate have been studied [51]. Among them, sodium carbonate and sodium bicarbonate (at 3% w/v) reduced up to 100% disease (*Penicillium* sp.) on inoculated fresh clementines and oranges [51]. Calcium carbonate maintained lower microbial loads in the FC melon samples after 8 days at 5°C [49]. However, little is known about the mode of action of these salts, although other possible mechanisms, apart from the high salt pH, like the induction of host defence responses might be involved [52].

2.6. Electrolyzed oxidizing water

Electrolyzed oxidizing water (EOW) is formed in an electrolysis chamber by the electrodialysis of a NaCl solution between an anode and a cathode [53]. Acidic EOW (pH 2.5–3.5; oxidation-reduction potential (ORP) 1000–1200 mV) is produced in the anode and alkaline EOW (pH 10–11.5; ORP −800 to −900 mV) is produced in the cathode. HCl, HOCl, Cl_2, OCl^-, and O_2 are formed in the anode, while the cathode produces hydroxyl ions, which can react with Na ions generating NaOH. EOW contains free chlorine as the main microbial inactivation agent showing higher microbicidal effect against pathogens and spoilage microorganisms than NaOCl [16, 54, 55]. EOW is considered as an eco-friendly technology that presents the following advantages over other sanitizing methods: easy-to-find and cheap materials (NaCl and water), simple and on-site production, and low operational expenses and trihalomethanes formation. Additionally, some cell electrodes, like boron-doped diamond electrodes, are able to oxidize organic matter, reducing then the environmental footprint of wastewater from fresh produce industry [56]. Nevertheless, EOW has a short shelf life, in some cases, being necessary to be produced on-site and recommended in a ventilated area due to the Cl_2 and H_2 production [5]. EOW has been approved at maximum concentration of 200 ppm by the FDA [57]. Neutral EOW (pH 7; ORP 700 mV) can be produced by the mixture of acidic and alkaline EOW [58]. The additional advantages from using neutral EOW are that it does not affect surface color, general appearance, or pH of FC vegetables [54].

EOW has been used in several works as an excellent disinfestation method of food equipment surfaces and tools reaching up to 9 log CFU cm^{-2} reductions for several pathogen biofilms like *L. monocytogenes, E. coli, Pseudomonas aeruginosa,* and *Staphylococcus aureus* [59–61].

The EOW (15–50 ppm free chlorine) was early studied on FC carrots, spinach, bell pepper, potato, and cucumber being considered as an effective disinfectant able to reduce microbial loads by 0.6–2.6 log units without product discoloration [54]. Neutral EOW (100 ppm free Cl; pH 7) achieved 0.5, 1.3, and >2.1 log mesophilic, psychrophilic, and yeast and mold reductions, respectively, in FC kailan-hybrid broccoli, showing similar microbial loads and good sensory quality to NaOCl-treated (100 ppm) samples after 19 days at 5°C [62]. Neutral EOW also showed similar microbial effectiveness to NaOCl to reduce natural microflora of FC lettuce with no impact on its physical and sensory quality [63–65]. Acidic and neutral EOW treatments at 70 and 100 ppm free Cl were studied on two broccoli varieties showing neutral EOW (100 ppm) the best microbial reductions after shelf life comparing to NaOCl (100 ppm) [58]. Furthermore, EOW-treated samples showed higher (up to 30%) total phenolic content and more stabilized myrosinase activity (the enzyme responsible for the formation of the bioactive isothiocyanates (ITC) in broccoli) than NaOCl-treated samples after shelf life [58]. Microbial reductions of 1–2 log units were observed in FC mizuna baby leaves treated with acidic EOW (40–100 ppm free Cl) and neutral EOW (40–100 ppm free Cl), with similar microbial effectiveness to NaOCl 100 ppm, showing neutral EOW better bacteriostatic effects than acidic EOW in some cases [66]. The sensory quality, physical structure, and health-promoting compounds of EOW-treated FC mizuna baby leaves were not significantly affected [66]. Neutral EOW (100 ppm free Cl) treatment reduced counts of inoculated *E. coli* and *S. enteritidis* in FC kailan-hybrid broccoli by approximately 2.6 log CFU g^{-1}. Nevertheless, the effectiveness of the last treatment was not increased when it was combined with ultraviolet (UV)–C (7.5 kJ m^{-2}) treatment. Similarly, neutral EOW treatment (306 ppm free Cl) of romaine lettuce reduced inoculated *E. coli* O157:H7, *S. typhimurium,* and *L. monocytogenes* loads by 2.0 log CFU g^{-1} [67]. *E. coli* O157:H7 grew slower in FC lettuce treated with NEW (50 ppm free Cl) during storage at 13–16°C, while no microbial growth was observed if the product was stored at ≤8°C [68].

The use of different organic and inorganic salts has been studied to increase electrolysis efficacy and avoid the corrosive effects of NaCl on equipment. Particularly, electrolyzed sodium bicarbonate allowed to control postharvest citrus rots as a result of direct inhibition and the induction of fruit resistance-response mechanisms showing normal electrolyzed water a less marked effect [69].

2.7. Ozone

Ozone (O$_3$) is a colorless gas with a pungent odor having an oxidizing potential (+2.07 V) 1.5 times higher than that of chlorine, which oxidizes the cell components of the microbial cell wall [70]. Ozone half-life is very short, from seconds to hours depending on temperature and water quality [71]. Thus, O$_3$ is commercially generated on-site by submitting oxygen, or air, to ultraviolet radiation (285 nm) or through an electrical charge leading to the cleavage of oxygen molecules to form ozone [5, 72].

Ozone solubility is 12 times lower than that of NaOCl. Nevertheless, ozone concentrations as low as 1–5 ppm are enough to reach good antimicrobial reductions. Nevertheless, higher O_3 concentrations are needed when it is applied as a gas treatment since its penetration into the cells, to achieve the disinfection effect, is affected by the humidity of the air [40, 73]. The O_3 effectiveness may be increased at lower pH (more stable) and temperature (higher residual ozone concentrations), higher relative humidity of the storage room (increasing its solubility on the moisture present on produce surfaces), and purity of the water (ozone is consumed by the matrix components reducing its efficacy) [74]. Ozone is spontaneously decomposed to the non-toxic O_2 when applied. Accordingly, O_3 has been approved as a GRAS product by the FDA to be used in the food industry [75]. However, O_3 can cause irritation to eyes and throat (at concentration higher than 0.2 ppm) of plant operators, is highly corrosive to the equipment, and the physicochemical properties of treated produce may be altered [5].

An ozonated water treatment at 0.4 ppm for 3 min has been recommended to maintain microbiological quality and firmness of tomato slices while it did not affect the physicochemical quality and organic acid contents [76]. The levels of *Salmonella* spp. inoculated in melons were reduced between 4.2 and 4.8 log CFU/rind-disk (12 cm^2) after a gaseous ozone treatment (10,000 ppm for 30 min under vacuum system) [77]. Counts of inoculated *E. coli* O157:H7 in spinach leaves were also decreased under a novel gaseous system capable of generating O_3 inside the sealed package at various geometries [78]. Respiration rate and browning of FC celery were reduced, while sensory quality was well maintained using ozonated water at 0.18 ppm for 5 min [79]. Ozonated water at 1 ppm reduced both enzyme activity and enzymatic browning of shredded lettuce [80]. Nevertheless, the latter enzyme inactivation showed a negative effect, as the reduction in activity of the texture-related pectin methyl esterase was correlated with a lower crispiness. FC rocket treated with ozonated water at 5 ppm for 10 min showed better sensory scores and microbial quality (psychrophilic and yeast and molds) than untreated samples after 12 days at 5°C, while total chlorophylls and carotenoids contents were unchanged [81]. Mesophilic, psychrotrophic, and yeast counts of FC tomato slices treated with ozonated water (3.8 ppm for 3 min) showed 1.9, 1.6, and 0.7 lower log units, respectively, than untreated samples after 10 days at 5°C [82]. The enzymatic antioxidant system of FC green peppers treated with gaseous ozone (6.42 mg cm^{-3} for 15 min) was induced during storage at 5°C, while polyphenol oxidase activity was reduced. A synergistic effect was even observed when the latter ozone treatment was combined with modified atmospheric packaging (3% O_2, 4% CO_2, and 93% N_2) of green peppers [83].

Nevertheless, O_3 seems to be not always highly effective. In that sense, low (<0.5 log units) inhibitory effect on mesophilic, psychrophilic, enterobacteria, molds, and LAB loads of FC "Galia" melon washed with ozonated water (0.4 ppm, up to 5 min) was observed after 10 days at 5°C, registering even higher yeast and molds compared to untreated samples [84]. However, FC 'Galia' melon washed for just 1 min with 80 ppm PAA showed the lowest microbial loads after 10 days at 5°C. However, the latter PAA antimocrobial effect was reduced when it was combined with ozonated water treatment. On the other side, FC "Galia" melon samples treated with ozonated water reduced respiration rates, while sugar contents and vitamin C were better maintained [84]. Shredded "Iceberg" lettuce treated with ozonated water (1 ppm, 120 s) showed lower sensory and microbiological quality than samples treated with NaOCl (200 ppm, 120 s) [85].

2.8. Essential oils

High antimicrobial properties have been studied with several plant essential oils (EOs) [86]. Generally, EOs possessing the strongest antibacterial properties are those that contain phenolic compounds such as carvacrol, eugenol, and thymol [87, 88]. In general, the mechanism of action of EOs against microorganisms involves the interaction of phenolic compounds with the proteins (porins) in the cytoplasmic membrane that can precipitate and lead to ions leakage of and other cell contents causing cell lysis [89]. Carvacrol solutions have shown good antimicrobial effects in different FC fruit and vegetables like lettuce, kiwifruit, apples, and melons treated with carvacrol-containing washing solutions [90–92]. Thymol (from thyme and oregano) and eugenol (from clove, *Syzygium* spp.) have also shown high antimicrobial and antioxidant effects on MAP-stored table grapes [93].

The EOs concentrations should be increased when tested *in vivo* in order to reach the same effectiveness than that observed *in vitro*. However, such high EOs concentrations may lead to EOs-related off-flavors transmission to the product. Furthermore, the lipophilic nature of EOs difficult their solution in the water-based washing solutions [86]. Therefore, the reduction of the EOs particle size (<100 nm) has been proposed as an alternative to improve EO antimicrobial efficiency through two important targets: (i) the possibility of enhancing physicochemical properties and stability; and (ii) the ability of improving biological activity of lipophilic compounds by increasing the surface area per unit of mass [94]. The antimicrobial and physical properties of different EOs (lemongrass, clove, tea tree, thyme, geranium, marjoram, palmarosa, rosewood, sage, or mint) have been reported to be enhanced when they were processed to nanoemulsions [94]. Such nanotechnology has been recently applied in FC carrots using nanoencapsulated carvacrol particles' incorporation in a washing solution, which reduced the characteristic off-flavors and EOs oxidation while keeping good microbial quality of the product during shelf life [95]. Furthermore, EOs may be included as antimicrobial agents in packaging films as reported of FC vegetables as proposed in a study using a carvacrol–polylactic acid film to inhibit *E. coli* ATCC 8739, *Fusarium oxysporum*, *Geotrichum candidum*, and *Phytophthora* spp. [96].

2.9. Isothiocyanates

Isothiocyanates (ITC) are sulfur compounds that can be formed in the *Brassica* vegetables after hydrolysis of glucosinolates by plant myrosinase. Antimicrobial activity of ITC has been reported for a wide range of foodborne microorganisms [97, 98]. The antimicrobial mechanism of ITC is not still clear, although it is hypothesized to be owed to the electrophilic nature of the central carbon atom located in the N=C=S group [99]. Among them, allyl-isothiocyanate has also shown high antimicrobial activity being listed as a GRAS [100]. FC lettuce treated with a washing solution of 75 ppm of benzyl-isothiocyanate (5 min) achieved a complete removal of total bacteria and inoculated *Salmonella* in the wash water, which proved to persist such antimicrobial effects in the processed water up to 48 h [101]. However, the low solubility of ITC highly limits its application as a sanitizing water treatment in the FR industry. Accordingly, integration of ITC in edible coatings of FC produce has been proposed with a longer antimicrobial effect due to slower release from the coating [96].

3. Biological-based methods

3.1. Bacteriocins

Bacteriocins are toxins of protein nature that are synthetized by bacteria to inhibit the microbial growth of similar or closely related bacterial strains. The solubility of bacteriocins may increase at lower pH, facilitating diffusion of bacteriocin molecules [102]. Nisin is a food-grade bacteriocin produced by *Lactococcus lactis* that is widely used in the food industry. The antimicrobial activity of this bacteriocin is owed to its action on the cell membrane forming pores leading to the microbial cell death [103]. Nisin is principally effective against Gram-positive bacteria, while it is not active against Gram-negative bacteria due to their outer membrane [104]. However, the nisin efficacy may be increased also due to Gram-negative bacteria using chelating agents (e.g., ethylene diaminotetracetic acid [EDTA]), acids, or osmotic shock, with the outer membrane destabilized before the nisin application [103]. Attending to natural microflora, a mesophilic reduction of approximately 2 log CFU g^{-1} was reached in FC "Galia" melon after a nisin (0.250 g L^{-1}) treatment combined with EDTA (0.100 g L^{-1}) and citric acid (2.0 g L^{-1}) [105]. Nisin and other bacteriocins (pediocin, coagulin, plantaricin C, and lacticin 481) were also tested in FC lettuce inoculated with *L. monocytogenes*, inducing nisin and coagulin a reduction of pathogen viability by 1.2–1.6 log units [106]. The inclusion of nisin (IU mL^{-1}) in a pectin coating applied on FC "Rojo Brillante" persimmon completely inhibited the growth of mesophilic bacteria and inoculated *E. coli, S. enteritidis*, and *L. monocytogenes* [107]. Nisin treatment (0.03% for 10 min) controlled microbial growth and maintained quality of Chinese yam during storage at 4°C [108]. Bacteriocin RUC9 produced by *L. lactis* reduced by 2.7 log units, the *L. monocytogenes* loads inoculated in FC lettuce, while nisin only achieved a pathogen reduction of 1.9 log units [109].

3.2. Biological control

The use of biological preservation (strains of *Enterobacteriaceae*, lactic acid bacteria, yeasts, and molds) has also been studied in several products like lettuce, apples, peaches, and strawberries [110–113]. Recently, the application of the strain CPA-7 of *Pseudomonas graminis*, isolated from apples, could reduce the foodborne pathogens on FC apples, peaches, and melon [114, 115]. Such antimicrobial effects of these antagonists may be explained by the triggered activity of defence-related enzymes [116].

3.3. Bacteriophages

Bacteriophages are viruses that infect and replicate in a bacteria causing their lysis and death [117]. Bacteriophages were earlier studied on FC fruit like apples and melons [118]. *S. typhimurium* and *S. enteritidis* inoculated in lettuce were highly reduced by 3.9 and 2.2 log units, respectively, using different lytic bacteriophages treatments for 60 min at room temperature [119]. Nevertheless, bacteriophages treatments should be optimized due to the impractical application in the FC industry. Accordingly, integration of bacteriophages in edible coatings of FC produce has been proposed as an effective antimicrobial coating in tomatoes, with such phages being stable up to 1 week at 4°C [120].

4. Physical-based treatments

4.1. Mild heat treatments

The use of mild heat treatments is a promising sanitizing technique for the FC industry, which may extend the FC product shelf life through microbial destruction and partial enzymatic inactivation. However, the treatment temperature and exposure time should be carefully selected for each product due to possible undesirable changes of sensory and nutritional quality. Fruit and vegetables treated by heat treatments within the range 40–60°C for 1–5 min, depending of the commodity, are still considered as a fresh product as it is defined [17, 121].

Hot water and vapor treatments have been studied in several fresh-cut commodities like pomegranate arils [122, 123], kiwifruit [124], lemons [125], peaches [126], apples [127], sunchoke [128], lettuce [46, 129, 130], rocket [131], spinach [132, 133], celery [134], eggplants [135], and onions [136]. PPO activity of fresh-cut pomegranate arils was 1.3-fold reduced by a hot water treatment at 55°C for 30 s, while total anthocyanins contents were similar to those of untreated samples after 7 days at 5°C [137].

Short vapor treatments, usually up to 15 s, usually steam jet-injection systems, have been also used as an alternative to hot water treatments due to less impact on sensory quality of the FC product. Accordingly, vapor treatment (95°C; 7–10 s) kept better quality of FC pomegranate arils [138] compared to the hot water (55°C; 30 s) treatment [139] during storage at 5°C, increasing the shelf life from 7 to 18 days. Furthermore, vapor heat treatments reduced up to 2-fold the total antioxidant capacity losses observed in fresh-cut pomegranate arils sanitized by conventional NaOCl treatment during shelf life [138]. Steam jet-injection treatment of fresh-cut lettuce for 10 s reduced respiration rate, partially inactivated browning-related enzymes and kept the mesophilic load as low as with a conventional NaOCl treatment [46]. Nevertheless, further research is needed to optimize the exposure conditions for FC commodities.

Heat treatments by microwave (750 W for 45–60 s) have been recently proposed to reduce natural microflora of FC carrots, which also prevented whitening and surface drying of samples during storage up to 7 days at 5°C [140]. The increased microbial growth due to plant cell disruption after the heat treatment may be controlled with the use of combined storage under modified atmosphere packaging.

4.2. UV radiation

Ultraviolet (UV) light is an electromagnetic radiation divided in four groups: UV-A, UV-B, UV-C, and vacuum UV [141]. UV-C radiation (λ = 190–280 nm) is a promising sanitizing technology for FC products, which offers several advantages: it does not leave any residue, no legal restrictions, easy to use, and it does not require extensive safety equipment to be implemented [142, 143]. UV-C is a non-ionizing radiation, which means it is an electromagnetic radiation that does not carry enough energy/quanta to ionize atoms or molecules and is represented mainly by visible light, UV rays, microwaves, and infrared. UV-C radiation in the range 250–260 nm is lethal to most microorganisms, including bacteria, viruses, protozoa, mycelial fungi, yeasts, and algae, showing the maxima germicidal effectiveness at 254 nm [144]. UV-C germicidal effect is based on the ability of this radiation to alter microbial DNA through dimer formation [142]. If the damage goes unrepaired, the accumulation of DNA photoproducts can be lethal to cells through the blockage of DNA replication and RNA transcription, which ultimately result in reproductive cell death. Nevertheless, it has been also stated that UV-C may lead to the conversion of bacteria in the viable but non-cultivable state as a strategy of protection against the UV-C germicidal effect (to economize on energy, induction of repair mechanisms, inhibit the generation of mutant bacteria, etc.) [145]. UV-C is a superficial sanitizing treatment with low penetration in the plant tissue as observed in carrot tissue where a transmittance below 20% was observed in a 0.1-mm layer of the carrot epidermis [146]. Accordingly, a UV-C treatment of 0.4 kJ m^{-2} applied to iceberg lettuce internally inoculated (using vacuum system) with *Salmonella* Montevideo P2 did not achieve significant pathogen reduction, while the same UV-C dose achieved a 2-log CFU g^{-1} reduction on the surface-inoculated lettuce [147]. UV-C effectiveness appears to be dependent on the treatment temperature, distance between sample and lamp, direction of lamp, UV intensity, and exposure time [148, 149]. Cell permeability may be changed with UV-C, depending on the tissue and UV dose, leading to increase of electrolytes, amino acids, and carbohydrates leakage, which can enhance the microbial growth [150]. Accordingly, the crucial point is to apply an appropriate UV-C dose that achieves the maximum microbial reduction without damaging the product.

The UV dose (D; usually expressed in kJ m^{-2}) is directly proportional to the product of UV intensity (I; usually expressed in W m^{-2}) and exposure time (t) according to the equation: $D = I \times t$. The three pathogens regulated for FC products (according to the European Regulation [151]), *E. coli*, *Salmonella* spp., and *L. monocytogenes*, were inoculated in the kailan-hybrid broccoli, and the inactivation rates with UV-C doses up to 15 kJ m^{-2} were modeled [149]. The inactivation curves showed a pronounced tailing effect achieving a UV-C dose of 2.5 kJ m^{-2} *E. coli*, *S. enteritidis*, and *L. monocytogenes* reductions of 1.22, 2.61, and 0.72 log cycles, respectively. Hence, *S. enteritidis* was the most sensitive microorganism, while *L. monocytogenes* was the most resistant. Doses higher than 2.5 kJ m^{-2} led to further additional inactivation in the inoculated kailan-hybrid broccoli, although the most important inactivating effect was achieved in the range 0–2.5 kJ m^{-2} as similarly found in other FC products [149, 152].

The UV-C effectiveness on natural microflora has also been studied in several FC fruit and vegetables. UV-C treatments (0.49, 4.9, and 9.8 kJ m^{-2}) of FC zucchini slices reduced microbial activity and deterioration during subsequent storage at 5 or 10°C [153]. Mesophilic loads of FC pomegranate arils were reduced by approximately 1 log units after 4.5 kJ m^{-2}, while yeasts were reduced >1.8 log units [139]. Similarly, UV-C radiation doses (4.5–9 kJ m^{-2}) of the kalian-hybrid broccoli reduced mesophilic loads by approximately 1.2 log units while enterobacteria and psychrophilic were unaffected [62, 154]. However, combination of UV-C with NEW in kalian-hybrid broccoli or with hot water (55°C for 30 s) in FC pomegranate arils did not achieve further microbial inactivations [62, 139]. FC tomatoes treated with a UV-C dose of 4 kJ m^{-2} and stored for 21 days at 12°C under MAP (5 kPa O$_2$ +1 kPa CO$_2$) retarded ripening and maintained better firmness and sensory attributes than UV-C treated samples stored under air conditions [155]. The range of 0–2.5 kJ m^{-2} UV-C dose achieved the most important mesophilic reductions in treated date palm [156]. The UV-C sanitizing effect has been studied in a wide FC products such as tomato [157], strawberry [158], watermelon [159], potatoes [160], and lettuce [161], among others.

Increases in bioactive compounds of several fruit and vegetables after UV treatments have been reported. Such enhancements of health-promoting compounds have been reported to be a consequence of the free radicals generated during irradiation that might act as stress signals and trigger stress responses leading to the observed bioactives increments [162]. Broccoli exposed to several UV-C doses (1.5–15 kJ m^{-2}) registered increases in its polyphenols content (up to 25%) after 19 days at 5°C [154]. FC tomatoes UV-C treated (0.97 kJ m^{-2}) showed higher total phenolic content than other samples treated with hot water (40°C, 30 min), ultrasounds (45 kHz; power of 80%; 30 min), or its combination, after 30 days at 10°C [157]. The lycopene content of watermelon was preserved with a UV-C dose of 2.8 kJ m^{-2}, although a lower dose of 1.6 kJ m^{-2} did not show the same benefit [159]. However, immediate bioactive increments after UV-C have been observed in several FC fruit and vegetables, being probably attributed to an enhanced compound extraction due to plant cell disruption as a consequence of the UV radiation. Therefore, phenolic compounds and flavonoid contents of FC mangoes were increased after UV-C doses of 2.46 and 4.93 kJ m^{-2} [163]. Hydroxycinnamoyl acid derivatives of FC broccoli were also increased by 4.5–4.8-fold after UV-C treatments (4.5–9.0 kJ m^{-2}) [154]. Total phenol and flavonoid contents of banana and guava were enhanced after UV-C treatment [164]. FC carrot treated with UV-C (9 kJ m^{-2}) showed higher chlorogenic content on total antioxidant capacity than untreated samples on processing day [146]. Then, besides the interest of UV-C radiation as a microbial safety method, this non-ionizing radiation may also be used as a tool to enhance or better preserve the health-promoting compounds of plant products during shelf life [165].

Application of UV-B radiation (280–320 nm) has also been proposed as a friendly and cheap nonmolecular tool to enhance the phenolic compounds in carrots and other horticultural crops during postharvest life [166–168]. FC carrot shreds treated with a UV-B dose of 1.5 kJ m^{-2} showed 23% higher total phenolic content (mainly chlorogenic acid) than untreated samples after 72 h at 15°C [169]. UV-A has also been reported to induce biosynthesis of anthocyanins in cherries [170], although its effects has not been widely reported, and further research must be conducted.

Low-light conditions during storage have been recently proposed as an innovative and eco-friendly postharvest technique to highly prolong the shelf life of FC products. Accordingly, the shelf life of FC lettuce (butterhead and iceberg) was highly extended when it was stored under low-light conditions ($\approx$5 µmol m^{-2} s^{-1} PAR; using either fluorescent or LED light) compared to samples stored under dark conditions [171]. Thus, lighting delayed cut-edge browning, reduced ascorbic acid degradation while carbohydrates levels were highly increased, although light samples did not show net photosynthesis according to photosynthetic activity measurements. Latter authors hypothesized that the observed prolonged shelf life in lit samples could be due to the higher levels of sugar and ascorbate that may act as antioxidants, may maintain membrane integrity, and may supply enough respiratory substrate to prevent ATP depletion.

4.3. Pulsed light

Pulsed light (PL) is a preservation technology that involves the use of intense short duration (1 μs–0.1 s) pulses of polychromatic light from UV to near infrared (100–1100 nm) emitted by an inert gas (e.g., xenon) lamp [172]. The microbicidal action of PL has been attributed to different mechanisms: photochemical, photothermal, and photophysical [173]. Several studies have shown the effects of PL treatments on inoculated microorganisms, native microflora, and quality aspects of FC spinach, lettuce, cabbage, carrot, mushrooms, avocado, and watermelon [174–179]. Microbial reductions up to 2.2 log units have been reported in different products such as lettuce, celery, spinach, bean sprouts, white cabbage, and green bell pepper, with the different antimicrobial effectiveness in different produce dependent on the location of microorganisms as well as the presence of protective substances present in the product [179, 180]. Further investigation may cover the reported microbial photoreactivation after PL treatments [181–184] and the PL efficiency due to shadow effects that is one of the main industrial limitations of PL technology.

4.4. Pulsed electric fields

Pulsed electric fields (PEFs) are based on the application of DC voltages for very short periods of time, usually μs, to the food material which is placed between two electrodes. PEF equipment consists of a treatment chamber, pulse generator, control system, data acquisition, and material-handling equipment [185]. PEF has been successfully applied for microbial inactivation in liquid food systems, although its application in plant tissues is limited due to the PEF-related plant cell disruption processes known as membrane breakdown, membrane permeabilization, or electroporation of the membrane [186]. Accordingly, PEF treatment (wave bipolar pulses at 2 kV cm^{-1} electric field strength, 1 μs pulse width, and 100 pulses s^{-1}) applied to blueberries (immersed in a saline fluid for PEF transmission) for just 2 min achieved 1 and 2 log unit reductions of inoculated *E. coli* and *L. innocua*, respectively [187]. Latter authors reported no PEF effects on color and appearance of blueberries, and the nutritional quality even enhanced, although PEF caused fruit softening. However, high electric field strength (333 V cm^{-1}) applied on a single pulse has been reported to not alter structure-related properties of FC onions, while such undesirable effects were observed using several pulses ($n \geq 10$) [185]. Accordingly, PEF is a promising technology to be used in FC produce, although the tissue changes as a function of the electrical field strength, and the number of pulses for each plant produce must be further investigated.

4.5. Cold plasma

Plasma is generated when an inert gas is in contact with electricity and is being considered as the fourth state of matter. Plasma is composed by charged particles, excited molecules, reactive species, and UV photons which induce microbial inactivation [188]. Plasma is generally classified as cold (non-thermal) and thermal plasma. Thermal plasma generation requires temperature and high pressure with heavy electrons. Cold plasma is generated at temperatures of 30–60°C under atmospheric or vacuum requiring low energy [189]. Among the main advantages of cold plasma are lower cost operating temperature and water consumption, together with timely production of the acting agents and lack of residues during production when compared to thermal and chemical treatments [190–192]. Cold plasma can be generated using either of the following devices: resistive barrier discharge, dielectric barrier discharge, corona discharge, radio frequency discharge, glow discharge, and atmospheric pressure plasma jet [193]. Plasma jet may be considered as the fastest plasma generation method to achieve microbial inactivation (4.3 ± 6.5 min) [194]. The most widely used gas in the published research has been air, followed by pure Ar, mixtures of He/O_2 and Ar/O_2, and pure N_2 [194]. The plasma inactivation capacity depends on several factors like the type of technology used to generate the plasma, the voltage, the feed gas, the treatment time, the species, the direct or indirect exposure and the concentration of the tested microorganisms, and the structural characteristics of the produce [192]. Cold plasma has a high potential to be used in the industry for fruit and vegetables according to studies of the last few years as recently reviewed [194]. Generally, plasma treatments are able to achieve microbial reductions of 2.7 ± 1.4 log units, ranging from 1.5 ± 1.0 log units for bacilli and spores to 3.3 ± 1.6 Log for *Listeria* spp., with treatment times of 22.2 ± 7.5 min for bacilli and spores and 3.5 ± 3.8 min for *E. coli* spp. [194]. Cold plasma has been also reported to highly (42–89%) reduce enzymatic browning of FC apples and potatoes [195, 196]. Nevertheless, the bactericidal mechanisms of cold plasma are still unclear being dependent on lots of factors related to processing parameters, environmental elements, and microbial properties [197].

4.6. Ultrasounds

Ultrasounds (US) are sound waves with amplitude higher than the upper audible limit of human threshold (above 20 kHz) that generate cavitation bubbles [198]. The antimicrobial properties derived from US is based on the combination of mechanical (responsible for the disinfection action leading to detachment) and chemical energy (responsible for the free radicals formation leading to destruction), produced from the collapse of latter generated bubbles, which increase the cell membranes permeability [40]. Consequently, DNA modifications of microbial cells are formed due to formed hot spots (due to collapse), with high temperatures and pressure, and released free radicals [199]. The US treatment should always ensure that US pressure levels (70 dB at 20 kHz or 100 dB at ≥25 kHz) are not surpassed according to UK Health Protection Agency recommendations [200]. The effectiveness of microbial inactivation achieved with US is influenced by microbial cell shape (high resistance of coccus), size (bigger cells are less resistant), Gram type (negative are less resistant), and cellular metabolism (anaerobes are less resistant) [201]. Among treatment parameters, US effectiveness is influenced by fluid temperature (optimum at 60°C which may be reduced to 20°C to avoid losses of produce sensory quality without highly affecting microbial US inactivation), water hardness, and dissolved gases content [202]. US treatment (40 kHz, 50 W) for 5 min of strawberries initially reduced natural mesophilic microflora by approximately 1 log unit [203]. However, latter authors showed that such initial antimicrobial effect was lost after 5 days at 8°C since similar reduction logs (regarding unwashed samples) to sterile water-washed (5 min) samples were achieved. On the other hand, such antimicrobial effects were maintained up to 9 days at 8°C when US treatment was combined with PAA (40 ppm) even showing a synergistic effect on mesophiles and the highest initial reductions on inoculated *S. enterica* subsp. Enterica, with sensory, physicochemical, and nutritional quality highly maintained [203]. No statistical differences among the effects of different frequencies (25, 32, and 70 kHz; 10 min) on achieved log reductions (1.5 log) were observed in FC lettuce inoculated with *S. typhimurium* [202]. Inoculated *E. coli*, *S. enteritidis*, and *L. innocua* loads of FC lettuce were reduced by 2.3, 5.7, and 1.9 log units with a US treatment for 30 min at 37 kHz without high color changes, although sensory quality was not studied in a parallel experiment with non-inoculated samples [204]. However, the produce matrix is highly important for US effectiveness since lower pathogen inactivations were observed in different products [204, 205].

4.7. High-pressure processing

High-pressure processing (HPP) uses elevated pressures, with or without the addition of heat, also called high-hydrostatic pressure processing since water is the most used pressure-transmitting fluid [206]. HPP is a promising eco-friendly sanitation treatment that may have a potential application in the FC industry. HPP may reach very high microbial inactivations as reviewed [207] while maintaining, or even enhancing, sensory properties of food products like aroma and taste. However, although the texture of tissues of firm FC products with low amounts of entrapped air remains unaffected, HPP may induce some alterations like water-soaked appearance of the product [208]. Furthermore, HPP may either enhance (leading to enzymatic browning reactions due to loss of membrane permeability and sub-cellular compartmentalization) or inhibit the activity of enzymes related to cell wall degradation in FC products [209]. HPP treatment (200–400 MPa, 3 min, 25°C) of FC persimmon induced changes in physicochemical quality (electrolyte leakage, texture, total soluble solids, pH, and color), which were a function of the amount of applied pressure compromising the consumer acceptance of the product [210]. However, latter authors reported that HPP may improve carotenoid extractability and tannin polymerization of FC persimmon, which could enhance its functionality and eliminate astringency, respectively. Plant cell membranes of FC peaches have been tried to be stabilized prior to HPP treatment (200 MPa for 10 min; 23–28°C) through penetration of Ca^{2+} into the plasma membrane using calcium chloride or calcium lactate soaking treatments (1–2% w/v for 5 min). Nevertheless, latter authors reported that loss of cell integrity due to HPP was not avoided with the calcium soakings probably due to a low Ca^{2+} penetration into the tonoplast membrane being recommended for future research a higher calcium concentration and/or improved Ca^{2+} impregnation (e.g., using vacuum infusion). Higher pressure levels (585 MPa) were more effective to inactivate enzymes and to preserve color of FC peaches than longer times being optimized a HPP treatment of 585 MPa for 1 min [211]. Nevertheless, due to the known baroresistance of some enzymes, like polyphenol oxidase (PPO), browning reactions may not be completely avoided during FC product shelf life, although MAP could limit such enzymatic activities due to low oxygen levels.

High-pressure carbon dioxide (HPCD) treatment has been proposed as another antimicrobial method being applied in FC carrots at 12 MPa (40°C) for 15 min and leading to complete inactivation of natural microflora being maintained after 28 days at 4°C together with the enzymatic stability [212]. HPCD treatment (12 MPa; 40°C; 20 min) of inoculated FC coconut also achieved *S. typhimurium* reductions of 4 log units, which was even enhanced to 8 log units when HPCD was combined with a high-power ultrasound treatment (10 W delivered every 2 min of treatment) [213].

5. Packaging under non-conventional gas mixtures

Modified atmosphere packaging (MAP) is a postharvest preservation technique based on the packaging of a perishable product within an atmosphere that has been modified compared with air conditions. There are two types of MAP: active or passive. Active MAP consists in the replacement on the initial present gases by a desired mixture. Passive MAP is progressively generated as a result of respiration of the product and gas transfer through the film, which has a selected permeability to gases, until the desired gas equilibrium atmosphere is reached.

The use of superatmospheric O_2 concentrations (>75 kPa O_2) during modified atmosphere storage (HO—high oxygen conditions) reduces aerobic and anaerobic microbial growth, prevents anaerobic fermentation, avoids non-desirable flavor changes, and inhibits enzymatic browning. The microbial toxicity to HO may be explained due to the unfavorable effects on the oxidation-reduction potential of the system, the oxidation of enzymes having sulfhydryl groups or disulfide bridges, and the accumulation of toxic reactive O_2 species [214]. HO-controlled atmosphere (75 kPa O_2 balanced with N_2) inhibited the mesophilic count of FC lettuce during storage for 10 days at 7°C [215, 216]. Chinese bayberries, strawberries, and blueberries stored under HO-controlled atmospheres (60–100 O_2 kPa balanced with N_2) inhibited decay during storage at 5°C and subsequent 2 days at 20°C, while chemical parameters and surface color were only slightly affected compared to samples stored under air conditions [217].

High CO_2 levels (maximum limits depending on the produce due to generation of related off-flavors) have also shown antimicrobial effects which are even stronger at low temperature because of enhanced CO_2 solubility. Accordingly, controlled atmosphere with 15 kPa CO_2 + 5 kPa O_2 (balanced with N_2) showed similar inhibitory effect to high O_2 on mesophilic growth of FC lettuce than samples stored at 0 kPa CO_2 + 75 kPa O_2 (balanced with N_2) [216]. Interestingly, a combined high O_2/CO_2 effect may be obtained using active HO MAP. The latter beneficial gas conditions are a result of the produce respiratory activity that generates antimicrobial CO_2 levels, while O_2 is inevitably reduced due to a combined effect of respiration and film diffusivity processes. Hence, the addition of CO_2 is unnecessary when high O_2 atmospheres are injected during active HO MAP. Accordingly, kailan-hybrid broccoli stored under high O_2/CO_2 (initial O_2/CO_2 of 70/0.02 kPa changing to 50/30 after 19 days at 5°C) showed 2.8 log units lower natural microflora load safter 19 days at 5°C compared to samples stored under passive MAP conditions (1.5–3.0 kPa O_2 + 16–21 kPa CO_2) [62]. Similarly, active HO MAP of inoculated kailan-hybrid broccoli showed 1.4 and 2.3 lower *E. coli* and *S. enteritidis* log units, respectively, after 19 days at 5°C regarding samples stored at passive MAP conditions [16]. However, such beneficial effect of HO was not observed when FC kailan-hybrid broccoli was stored at 10°C. Yeast and mold growth in FC pomegranate arils packaged under active MAP (initial O_2/CO_2 of 70/0.02 kPa changing to 20/5.5 kPa after 14 days at 5°C) was highly inhibited (up to 1.2 log units inhibition) during storage for 14 days at 5°C [139]. Such beneficial high O_2/CO_2 effects on microbial growth of FC produce packaged under active HO MAP have also been observed in other studies [218, 219].

Enzymatic browning of FC produce has been shown to be reduced under HO atmospheres. Accordingly, active HO MAP (80 kPa O_2 + 20 kPa CO_2) delayed browning of FC lettuce during storage for 10 days at 5°C [220]. Active HO MAP (initial O_2/CO_2 of 80/0 kPa, balanced with N_2) of FC pomegranate arils reduced enzymatic browning related to PPO [123, 137], while formed off-odors were highly controlled with active high CO_2 MAP (20 kPa CO_2 balanced with N_2) [139]. Active HO MAP of FC celeriac, mushrooms, and chicory endives (initial O_2/CO_2 of 95/0 kPa, balanced with N_2, changing to 10–20/10–50 after 7 days at 4°C) were more effective to control enzymatic browning than low O_2 atmospheres [221]. It is hypothesized that high O_2 may cause substrate inhibition of PPO or high contents of colorless quinones formed cause feedback PPO inhibition. High O_2 levels kept the initial color and firmness of fresh-cut melon retarding anaerobic fermentation better than low O_2 atmospheres [222]. Pre-treatment of whole "Spartan" apple with 100 kPa O_2 (up to 19 days at 1°C) before cutting decreased surface browning, flesh softening, and off-flavor in FC apple slices [223]. Such inhibition of enzymatic browning was related to retention of cellular integrity while in low O_2 pre-treatment (1 kPa O_2, balanced with N_2) would have another inhibitory browning mechanism on apple slices. Furthermore, the 100 kPa O_2 pre-treatment before slicing apples can reduce the dependence on antioxidant additives to inhibit slices browning. Sensory quality of produce stored under high O_2 atmospheres has been shown to be better than low O_2 atmospheres due to fermentation processes [218].

High O_2 levels are also considered as postharvest abiotic stresses able to increase PAL activity and consequently phenolic biosynthesis as observed in carrot shreds stored under high O_2-controlled atmosphere (80 kPa O_2 balanced with N_2) [146]. Furthermore, high O_2 and CO_2 levels may be tolerated by FC carrots maintaining their fresh characteristics and reducing microbial growth [224]. Such abiotic stress may be used as a tool to increase the health-promoting compounds of plant material to subsequently obtain functional beverages after correspondent thermal or non-thermal treatments to ensure microbial quality and safety [225].

The use of MAP under mixtures of non-conventional gases such as Ar, He, Xe, or N_2O has been proposed to maintain the quality of FC produce extending its shelf life. The latter gases may be chemically inert, but they have some physiological and/or antimicrobial properties, even though it does not seem to be through modification of enzyme activity [226]. Ar, He, or N_2 atmospheres mixed with low O_2 showed different diffusive properties, since Ar and He are monoatomic and smaller in size than N_2 [227]. Treatment of asparagus spears for 24 h at 4°C under an atmosphere of Ar and Xe at 2:9 (v:v) reduced RR and bract opening leading to a subsequent shelf life of 12 days at 4°C showing better quality than those samples treated with an atmosphere of 5 kPa O_2 + 5 kPa CO_2 [228]. Microbial quality and some bioactive compounds were highly preserved in FC red chard baby leaves stored under active He MAP (100%) during 8 days at 5°C [229]. N_2O has a direct effect on cell metabolism achieving shelf life extension of FC produce. It may be explained since N_2O has 77% solubility in fruit cell, while its absorption in tissues is completely reversible [230]. FC spinach leaves stored under active N_2O MAP (100%)

showed low microbial growth after 8 days at 5°C, with chlorophylls and phenolics being well preserved [231]. Different N_2O and N_2 combinations (including always 3% O_2) were used as active MAP for FC lettuce and wild rocket during storage up to 12 days at 5°C, suggesting such results that N_2O does not improve the produce quality compared to N_2 [232]. Lower microbial loads have also been observed in FC watercress and arugula leaves at the end of cold storage when active MAP containing N_2, Ar, He, Xe, or N_2O were used [233, 234].

6. Future research needs and conclusions

FC plant produce is greatly vulnerable to microbial spoilage, and cultivar selection is probably the most important factor in FC overall quality and shelf life. With the intention of better inhibition of microbial spoilage, and subsequently decrease in decay and safety problems, genetic cultivar selection should turn to retard ripening and senescence, low ethylene production and/or sensitivity, and enhanced firmness, well adapted to minimal processing and increased antioxidant systems. While chlorine is widely used by the FC industry to ensure safety, new eco-friendly techniques/technologies are needed to replace the latter chemical treatment due to the production of carcinogenic compounds. Several eco-friendly strategies such as ozone, UV-C light, natural antimicrobial substances, GRAS chemical, and biological compounds can decrease microbial loads of FC products and extend their shelf life. Other advanced techniques such as pulsed electric fields, ultrasounds, and high-pressure processing, among others, are promising sanitizing strategies for the FC industry due to high microbicidal rates. However, the treatment parameters of latter technologies to be applied in FC products need further research to be optimized for each specific commodity. Furthermore, the potential and limits of these innovative eco-friendly techniques must be well defined and included in the regulations. In that way, modeling tools to predict microbial inactivation and product shelf life are very useful, principally to optimize production and distribution. Application of nanotechnology to FC products is also a promising future in order to produce products with extended shelf life and excellent quality meeting always the food safety. Nevertheless, regulations related to nanoparticles' inclusion in food products need to be better defined by institutions. The combination of well-designed integrated production, handling, processing, and distribution chains for FC produces is crucial for achieving the high quality and safety demanded by consumers.

Acknowledgements

The authors are grateful to European Union, DG, for Science Research and Development (CRAFT Program Contracts SMT4-CT98-5530, QLK1-1999-707917, and OLK1-CT-2002-70791), 7th Framework Programme (FP7-BBBE.2013.1.2-02), Spanish Ministry for Education and Science and Spanish Ministry of Economy and Competitiveness (Projects ALI-95-0001, ALI-98-1006, AGL2005-08189-C02-01/ALI; AGL2007-63861/ALI; AGL2010-19201-C04-02-AGR; AGL2013-48830-C2-1-R), the Consejería de Educación y Cultura de la Comunidad

Autónoma de la Región de Murcia (SUE-OA 06/02-0008), and Fundación Séneca de la Región de Murcia in Spain (Project AGR/92, PS96/CA/d3, AGR/3/FS/02 and 00553/PI/04) for financial support. We also acknowledge the following companies for financing research and development contract with us: Repsol Petróleo S.A. (Madrid), Plásticos de Alzira, S.A., Sakata Seed Iberica S.L.U. (Valencia), Primaflor SAT (Almería), Canarihorta (Las Palmas de Gran Canaria), Frutas Hermanos Mira S.L. (Alicante), Kernel Export S.L., Frutas Esparza S.L., Perichán SAT, Pozosur S.L. (Murcia), and Sakata Seed (Valencia). The good skills and work of all researches whenever involved in the Postharvest and Refrigeration Group are very much appreciated.

Author details

Francisco Artés-Hernández[1]*, Ginés Benito Martínez-Hernández[1], Encarna Aguayo[1], Perla A. Gómez[2] and Francisco Artés[1]

*Address all correspondence to: fr.artes-hdez@upct.es

1 Postharvest and Refrigeration Group, Department of Food Engineering, Universidad Politécnica de Cartagena, Cartagena, Spain

2 Institute of Plant Biotechnology, Universidad Politécnica de Cartagena, Edificio I+D+i, Cartagena, Spain

References

[1] Slavin JL, Lloyd B. Health benefits of fruits and vegetables. Advances in Nutrition: An International Review Journal. 2012;**3**:506-516. DOI: 10.3945/an.112.002154

[2] Artés F, Allende A. Processing lines and alternative preservation techniques to prolong the shelf-life of minimally fresh processed leafy vegetables. European Journal of Horticultural Science. 2005;**70**:231-245

[3] Nieuwenhuijsen MJ, Toledano MB, Elliott P. Uptake of chlorination disinfection by-products; a review and a discussion of its implications for exposure assessment in epidemiological studies. Journal of Exposure Analysis and Environmental Epidemiology. 2000;**10**:586-599. DOI: 10.1038/sj.jea.7500139

[4] Suslow TV. Postharvest Chlorination- Basic Properties and Key points for Effective Disinfection. University of California: Division of Agriculture and Natural Resources; 1997. p. 8003

[5] Artés F, Gómez P, Aguayo E, Escalona V, Artés-Hernández F. Sustainable sanitation techniques for keeping quality and safety of fresh-cut plant commodities. Postharvest Biology and Technology. 2009;**51**:287-296. DOI: 10.1016/j.postharvbio.2008.10.003

[6]   European Union (EU). Commission implementing regulation (EU) 2016/672 approving peracetic acid as an existing active substance for use in biocidal products for product-types 11 and 12. Official Journal of the European Union. 2006;L116:3-7

[7]   Food and Drug Administration (FDA). Title 21: Food and Drugs. Section. Part 173: Secondary direct food additives permitted in food for human consumption [Internet]. 2016. Available from: https://www.accessdata.fda.gov/scripts/cdrh/cfdocs/cfcfr/CFR Search.cfm?CFRPart=173 [Accessed: 2017-07-10]

[8]   Stampi S, De Luca G, Zanetti F. Evaluation of the efficiency of peracetic acid in the disinfection of sewage effluents. Journal of Applied Microbiology. 2001;**91**:833-838. DOI: 10.1046/j.1365-2672.2001.01451.x

[9]   Park CM, Beuchat LR. Evaluation of sanitizers for killing *Escherichia coli* O157:H7, *Salmonella* and naturally occurring microorganisms on cantaloupes, honeydew melons, and asparagus. Dairy Food and Environmental Sanitation. 1999;**19**:842-847

[10] Wright JR, Summer SS, Hackney CR, Pierson MD, Zoecklein W. Reduction of *Escherichia coli* O157:H7 on apples using wash and chemical sanitizers treatments. Dairy Food and Environmental Sanitation. 2000;**20**:120-126

[11] Ruiz-Cruz S, Acedo-Félix E, Díaz-Cinco M, Islas-Osuna MA, González-Aguilar GA. Efficacy of sanitizers in reducing *Escherichia coli* O157:H7, *Salmonella* spp. and *Listeria monocytogenes* populations on fresh-cut carrots. Food Control. 2007;**18**:1383-1390. DOI: 10.1016/j.foodcont.2006.09.008

[12] Silveira A, Aguayo E, Leglise A, Artés F. Emerging sanitizers and clean room improved the microbial quality of fresh-cut 'Galia' melon. In: 3rd International Symposium on Food and Agricultural Products: Processing and Innovations; 24-26 September 2007; Naples. Milan: Associazione Italiana Di Ingegneria Chimica (AIDIC); 2007. CDrom

[13] Vandekinderen I, Devlieghere F, De Meulenaer B, Ragaert P, Van Camp J. Optimization and evaluation of a decontamination step with peroxyacetic acid for fresh-cut produce. Food Microbiology. 2009;**26**:882-888. DOI: 10.1016/j.fm.2009.06.004

[14] Ge C, Bohrerova Z, Lee J. Inactivation of internalized *Salmonella Typhimurium* in lettuce and green onion using ultraviolet C irradiation and chemical sanitizers. Journal of Applied Microbiology. 2013;**114**:1415-1424. DOI: 10.1111/jam.12154

[15] Neo SY, Lim PY, Phua LK, Khoo GH, Kim SJ, Lee SC, Yuk HG. Efficacy of chlorine and peroxyacetic acid on reduction of natural microflora, *Escherichia coli* O157:H7, *Listeria monocyotgenes* and *Salmonella* spp. on mung bean sprouts. Food Microbiology. 2013;**36**:475-480. DOI: 10.1016/j.fm.2013.05.001

[16] Martínez-Hernández GB, Navarro-Rico J, Gómez PA, Otón M, Artés F, Artés-Hernández F. Combined sustainable sanitising treatments to reduce *Escherichia coli* and *Salmonella enteritidis* growth on fresh-cut kailan-hybrid broccoli. Food Control. 2015;**47**:312-317. DOI: 10.1016/j.foodcont.2014.07.029

[17] Parish ME, Beuchat LR, Suslow TV, Harris LJ, Garrett EH, Farber JN, Busta FF. Methods to reduce/eliminate pathogens from fresh and fresh-cut produce. Comprehensive Reviews in Food Science and Food Safety. 2003;**2**:161-173. DOI: 10.1111/j.1541-4337.2003.tb00033.x

[18] Betts G, Everis L. Alternatives to hypochlorite washing systems for the decontamination of fresh fruit and vegetables. In: Jongen W, editor. Improving the Safety of Fresh Fruit and Vegetables. 1st ed. Wageningen: Woodhead Publishing Limited; 2005. p. 351-372. DOI: 10.1533/9781845690243.3.351I

[19] European Food Safety Authority (EFSA). Scientific opinion on the safety of gaseous chlorine dioxide as a preservative slowly released in cold storage areas. EFSA Journal. 2016;**14**:4388-4406. DOI: 10.2903/j.efsa.2016.4388

[20] Trinetta V, Vaidya N, Linton R, Morgan M. Evaluation of chlorine dioxide gas residues on selected food produce. Journal of Food Science. 2011;**76**:T11–T15. DOI: 10.1111/j.1750-3841.2010.01911.x

[21] United States Environmental Protection Agency (EPA). Chapter 4: Chlorine dioxide. In: EPA, editor. EPA Guidance Manual: Alternative Disinfectants and Oxidants. Office of Water (4607);1999. p. 36

[22] Keskinen LA, Burke A, Annous BA. Efficacy of chlorine, acidic electrolyzed water and aqueous chlorine dioxide solutions to decontaminate *Escherichia coli* O157:H7 from lettuce leaves. International Journal of Food Microbiology. 2009;**132**:134-140. DOI: 10.1016/j.ijfoodmicro.2009.04.006

[23] Mahmoud BSM, Bhagat AR, Linton RH. Inactivation kinetics of inoculated *Escherichia coli* O157:H7, *Listeria monocytogenes* and *Salmonella enterica* on strawberries by chlorine dioxide gas. Food Microbiology. 2007;**24**:736-744. DOI: 10.1016/j.fm.2007.03.006

[24] Mahmoud BSM, Linton RH. Inactivation kinetics of inoculated *Escherichia coli* O157:H7 and *Salmonella enterica* on lettuce by chlorine dioxide gas. Food Microbiology. 2008;**25**:244-252. DOI: 10.1016/j.fm.2007.10.015

[25] Sy KV, Murray MB, Harrison MD, Beuchat LR. Evaluation of gaseous chlorine dioxide as a sanitizer for killing *Salmonella*, *Escherichia coli* O157:H7, *Listeria monocytogenes*, and yeasts and molds on fresh and fresh-cut produce. Journal of Food Protection. 2005;**68**:1176-1187. DOI: 10.4315/0362-028X-68.6.1176

[26] Rodgers SL, Cash JN, Siddiq M, Ryser ET. A comparison of different chemical sanitizers for inactivating *Escherichia coli* O157:H7 and *Listeria monocytogenes* in solution and on apples, lettuce, strawberries, and cantaloupe. Journal of Food Protection. 2004;**67**:721-731

[27] Chung CC, Huang TC, Yu CH, Shen FY, Chen HH. Bactericidal effects of fresh-cut vegetables and fruits after subsequent washing with chlorine dioxide. International Proceedings of Chemical, Biological & Environmental Engineering. 2011;**9**:107-112

[28] Tomás-Callejas A, López-Gálvez F, Sbodio A, Artés F, Artés-Hernández F, Suslow TV. Chlorine dioxide and chlorine effectiveness to prevent *Escherichia coli* O157:H7 and *Salmonella* cross-contamination on fresh-cut Red Chard. Food Control. 2012;**23**:325-332. DOI: 10.1016/j.foodcont.2011.07.022

[29] López-Velasco G, Tomás-Callejas A, Sbodio A, Artés-Hernández F, Suslow TV. Chlorine dioxide dose, water quality and temperature affect the oxidative status of tomato processing water and its ability to inactivate *Salmonella*. Food Control. 2012;**26**:28-35. DOI: 10.1016/j.foodcont.2011.12.016

[30] Tomás-Callejas A, López-Velasco G, Artés F, Artés-Hernández F. Acidified sodium chlorite optimisation assessment to improve quality of fresh-cut tatsoi baby leaves. Journal of the Science of Food and Agriculture. 2012;**92**:877-885. DOI: 10.1002/jsfa.4664

[31] Ölmez H, Kretzschmar U. Potential alternative disinfection methods for organic fresh-cut industry for minimizing water consumption and environmental impact. LWT – Food Science and Technology. 2009;**42**:686-693. DOI: 10.1016/j.lwt.2008.08.001

[32] Sapers GM, Miller RL, Pilizota V, Kamp F. Shelf-life extension of fresh mushrooms (*Agaricus bisporus*) by application of hydrogen peroxide and browning inhibitors. Journal of Food Science. 2001;**66**:362-366. DOI: 10.1111/j.1365-2621.2001.tb11347.x

[33] Huang Y, Ye M, Chen H. Efficacy of washing with hydrogen peroxide followed by aerosolized antimicrobials as a novel sanitizing process to inactivate *Escherichia coli* O157:H7 on baby spinach. International Journal of Food Microbiology. 2012;**153**:306-313. DOI: 10.1016/j.ijfoodmicro.2011.11.018

[34] Ukuku DO, Fett W. Behavior of *Listeria monocytogenes* inoculated on cantaloupe surfaces and efficacy of washing treatments to reduce transfer from rind to fresh-cut pieces. Journal of Food Protection. 2002;**65**:924-930. DOI: 10.4315/0362-028X-65.6.924

[35] Ukuku DO, Mukhopadhyay S, Geveke D, Olanya M, Niemira B. Effect of hydrogen peroxide in combination with minimal thermal treatment for reducing bacterial populations on cantaloupe rind surfaces and transfer to fresh-cut pieces. Journal of Food Protection. 2016;**79**:1316-1324. DOI: 10.4315/0362-028x.jfp-16-046

[36] Sapers G. Hydrogen peroxide as an alternative to chlorine for sanitizing fruits and vegetables. IFIS Publishing-Food Science Central [Internet]. 2003. Available from: https://foodinfo.ifis.org/ [Accessed: 2017-07-10]

[37] Van Haute S, Tryland I, Veys A, Sampers I. Wash water disinfection of a full-scale leafy vegetables washing process with hydrogen peroxide and the use of a commercial metal ion mixture to improve disinfection efficiency. Food Control. 2015;50:173-183. DOI: 10.1016/j.foodcont.2014.08.028

[38] Lianou A, Koutsoumanis KP, Sofos JN. Organic acids and other chemical treatments for microbial decontamination of food. In: Demirci A, Ngadi MO, editors. Microbial Decontamination in the Food Industry. 1st ed. Cambridge: Woodhead Publishing; 2012. p. 592-664. DOI: 10.1533/9780857095756.3.592I

[39] Carpenter CE, Broadbent JR. External concentration of organic acid anions and pH: Key independent variables for studying how organic acids inhibit growth of bacteria in mildly acidic foods. Journal of Food Science. 2009;**74**:R12–R15. DOI: 10.1111/j.1750-3841.2008.00994.x

[40] Meireles A, Giaouris E, Simões M. Alternative disinfection methods to chlorine for use in the fresh-cut industry. Food Research International. 2016;**82**:71-85. DOI: 10.1016/j.foodres.2016.01.021

[41] Gurtler JB, Mai TL. Preservatives – Traditional preservatives – Organic acids. In: Batt CA Tortorello ML, editors. Encyclopedia of Food Microbiology. Oxford: Academic Press; 2014. pp. 119-130

[42] Aguayo E, Allende A, Artés F. Keeping quality and safety of minimally fresh processed melon. European Food Research and Technology. 2003;**216**:494-499. DOI: 10.1007/s00217-003-0682-7

[43] Gómez P, Artés F. Ascorbic and citric acids to preserve quality of minimally processed green celery. In: Proceedings of the IV Postharvest Iberian Symposium; 6-9 October 2004; Oeiras. Lisbon: APH; 2004. pp. 369-373

[44] Akbas MY, Olmez H. Inactivation of *Escherichia coli* and *Listeria monocytogenes* on iceberg lettuce by dip wash treatments with organic acids. Letters in Applied Microbiology. 2007;**44**:619-624. DOI: 10.1111/j.1472-765X.2007.02127.x

[45] Park SH, Choi MR, Park JW, Park KH, Chung MS, Ryu S, Kang DH. Use of organic acids to inactivate *Escherichia coli* O157:H7, *Salmonella typhimurium*, and *Listeria monocytogenes* on organic fresh apples and lettuce. Journal of Food Science. 2011;**76**:M293-M298. DOI: 10.1111/j.1750-3841.2011.02205.x

[46] Martín-Diana AB, Rico D, Frías J, Henehan GTM, Mulcahy J, Barat JM, Barry-Ryan C. Effect of calcium lactate and heat-shock on texture in fresh-cut lettuce during storage. Journal of Food Engineering. 2006;**77**:1069-1077. DOI: 10.1016/j.jfoodeng.2005.08.037

[47] Rico D, Martín-Diana AB, Frías JM, Barat JM, Henehan GTM, Barry-Ryan C. Improvement in texture using calcium lactate and heat-shock treatments for stored ready-to-eat carrots. Journal of Food Engineering. 2007;**79**:1196-1206. DOI: 10.1016/j.jfoodeng.2006.04.032

[48] Martín-Diana AB, Rico D, Barry-Ryan C, Frías JM, Mulcahy J, Henehan GTM. Comparison of calcium lactate with chlorine as a washing treatment for fresh-cut lettuce and carrots: Quality and nutritional parameters. Journal of the Science of Food and Agriculture. 2005;**85**:2260-2268. DOI: 10.1002/jsfa.2254

[49] Aguayo E, Escalona VH, Artés F. Effect of hot water treatment and various calcium salts on quality of fresh-cut 'Amarillo' melon. Postharvest Biology and Technology. 2008;**47**:397-406. DOI: 10.1016/j.postharvbio.2007.08.001

[50] Techakanon C, Barrett DM. The effect of calcium chloride and calcium lactate pretreatment concentration on peach cell integrity after high-pressure processing. International Journal of Food Science & Technology. 2017;**52**:635-643. DOI: 10.1111/ijfs.13316

[51] Youssef K, Ligorio A, Sanzani SM, Nigro F, Ippolito A. Control of storage diseases of citrus by pre- and postharvest application of salts. Postharvest Biology and Technology. 2012;**72**:57-63. DOI: 10.1016/j.postharvbio.2012.05.004

[52] Youssef K, Sanzani SM, Ligorio A, Ippolito A, Terry LA. Sodium carbonate and bicarbonate treatments induce resistance to postharvest green mould on citrus fruit. Postharvest Biology and Technology. 2014;**87**:61-69. DOI: 10.1016/j.postharvbio.2013.08.006

[53] Ongeng D, Devlieghere F, Debevere J, Coosemans J, Ryckeboer J. The efficacy of electrolysed oxidising water for inactivating spoilage microorganisms in process water and on minimally processed vegetables. International Journal of Food Microbiology. 2006;**109**:187-197. DOI: 10.1016/j.ijfoodmicro.2005.12.013

[54] Izumi H. Electrolyzed water as a disinfectant for fresh-cut vegetables. Journal of Food Science. 1999;**64**:536-539. DOI: 10.1111/j.1365-2621.1999.tb15079.x

[55] Ju S-Y, Ko J-J, Yoon H-S, Seon S-J, Yoon Y-R, Lee D-I, Kim S-Y, Chang H-J. Does electrolyzed water have different sanitizing effects than sodium hypochlorite on different vegetable types? British Food Journal. 2017;**119**:342-356. DOI:10.1108/BFJ-06-2016-0283

[56] Kapałka A, Fóti G, Comninellis C. Kinetic modelling of the electrochemical mineralization of organic pollutants for wastewater treatment. Journal of Applied Electrochemistry. 2008;**38**:7-16. DOI: 10.1007/s10800-007-9365-6

[57] Food and Drug Administration (FDA). Food Additive Status List. [Internet]. 2013. Available from: https://www.fda.gov/food/ingredientspackaginglabeling/foodadditivesingredients/ucm091048.htm [Accessed: 2017-07-10]

[58] Navarro-Rico J, Artés-Hernández F, Gómez PA, Núñez-Sánchez MÁ, Artés F, Martínez-Hernández GB. Neutral and acidic electrolysed water kept microbial quality and health promoting compounds of fresh-cut broccoli throughout shelf life. Innovative Food Science and Emerging Technologies. 2014;**21**:74-81. DOI: 10.1016/j.ifset.2013.11.004

[59] Arévalos-Sánchez M, Regalado C, Martín SE, Meas-Vong Y, Cadena-Moreno E, García-Almendárez BE. Effect of neutral electrolyzed water on lux-tagged *Listeria monocytogenes* EGDe biofilms adhered to stainless steel and visualization with destructive and non-destructive microscopy techniques. Food Control. 2013;**34**:472-477. DOI: 10.1016/j.foodcont.2013.05.021

[60] Kim C, Hung Y-C, Brackett RE, Frank JF. Inactivation of *Listeria monocytogenes* biofilms by electrolyzed oxidizing water. Journal of Food Processing and Preservation. 2001;**25**:91-100. DOI: 10.1111/j.1745-4549.2001.tb00446.x

[61] Deza MA, Araujo M, Garrido MJ. Inactivation of *Escherichia coli, Listeria monocytogenes, Pseudomonas aeruginosa* and *Staphylococcus aureus* on stainless steel and glass surfaces by neutral electrolysed water. Letters in Applied Microbiology. 2005;**40**:341-346. DOI: 10.1111/j.1472-765X.2005.01679.x

[62] Martínez-Hernández GB, Artés-Hernández F, Gómez PA, Formica AC, Artés F. Combination of electrolysed water, UV-C and superatmospheric O_2 packaging for improving fresh-cut broccoli quality. Postharvest Biology and Technology. 2013;**76**:125-134. DOI: 10.1016/j.postharvbio.2012.09.013

[63] Wang, H. Feng, H., Luo, Y. Surface treatment of fresh-cut lettuce with acidic electrolyzed water to extend shelf life. In: Proceedings of the Annual Meeting of the Institute of Food Technologist (IFT); 12-16 July 2003; Chicago IL. Chicago: IFT; 2003. p. 265

[64] Rico D, Martín-Diana AB, Barry-Ryan C, Frías JM, Henehan GTM, Barat JM. Use of neutral electrolysed water (EW) for quality maintenance and shelf-life extension of minimally processed lettuce. Innovative Food Science and Emerging Technologies. 2008;**9**:37-48. DOI: 10.1016/j.ifset.2007.05.002

[65] Koseki S, Itoh K. Prediction of microbial growth in fresh-cut vegetables treated with acidic electrolyzed water during storage under various temperature conditions. Journal of Food Protection. 2001;**64**:1935-1942. DOI: 10.4315/0362-028X-64.12.1935

[66] Tomás-Callejas A, Martínez-Hernández GB, Artés F, Artés-Hernández F. Neutral and acidic electrolyzed water as emergent sanitizers for fresh-cut mizuna baby leaves. Postharvest Biology and Technology. 2011;**59**:298-306. DOI: 10.1016/j.postharvbio.2010.09.013

[67] Yang H, Swem BL, Li Y. The effect of pH on inactivation of pathogenic bacteria on fresh-cut lettuce by dipping treatment with electrolyzed water. Journal of Food Science. 2003;**68**:1013-1017. DOI: 10.1111/j.1365-2621.2003.tb08280.x

[68] Posada-Izquierdo GD, Pérez-Rodríguez F, López-Gálvez F, Allende A, Gil MI, Zurera G. Modeling growth of *Escherichia coli* O157:H7 in fresh-cut lettuce treated with neutral electrolyzed water and under modified atmosphere packaging. International Journal of Food Microbiology. 2014;**177**:1-8. DOI: 10.1016/j.ijfoodmicro.2013.12.025

[69] Fallanaj F, Ippolito A, Ligorio A, Garganese F, Zavanella C, Sanzani SM. Electrolyzed sodium bicarbonate inhibits *Penicillium digitatum* and induces defence responses against green mould in citrus fruit. Postharvest Biology and Technology. 2016;**115**:18-29. DOI: 10.1016/j.postharvbio.2015.12.009

[70] Horvath ML, Bilitzky L, Huttner J. Fields of utilization of ozone. In: Clark RJH, editor. Ozone. New York (USA): Elsevier Science Publishing Co.; 1985. pp. 257-316

[71] Antoniou MG, Andersen HR. Evaluation of pretreatments for inhibiting bromate formation during ozonation. Environmental Technology. 2012;**33**:1747-1753. DOI: 10.1080/09593330.2011.644586

[72] Graham DM. Use of ozone for food processing. Food Technology. 1997;**51**:72-75. DOI: 10.1016/j.lwt.2003.10.014

[73] Horvitz S, Cantalejo MJ. Application of ozone for the postharvest treatment of fruits and vegetables. Critical Reviews in Food Science and Nutrition. 2014;**54**:312-339. DOI: 10.1080/10408398.2011.584353

[74] Tzortzakis N, Chrysargyris A. Postharvest ozone application for the preservation of fruits and vegetables. Food Reviews International. 2017;**33**:270-315. DOI: 10.1080/87559129.2016.1175015

[75] Food and Drug Administration (FDA). Title 21: Food and Drugs. Section. Part 173: Secondary direct food additives permitted in food for human consumption [Internet].

2016. Available from: https://www.accessdata.fda.gov/scripts/cdrh/cfdocs/cfcfr/CFRSearch. cfm?fr=173.368 [Accessed: 2017-07-10]

[76] Aguayo E, Escalona V, Silveira AC, Artés F. Quality of tomato slices disinfected with ozonated water. Food Science and Technology International. 2014;**20**:227-235. DOI: 10.1177/1082013213482846

[77] Selma MV, Ibanez AM, Cantwell M, Suslow T. Reduction by gaseous ozone of *Salmonella* and microbial flora associated with fresh-cut cantaloupe. Food Microbiology. 2008;**25**:558-565. DOI: 10.1016/j.fm.2008.02.006

[78] Klockow PA, Keener KM. Safety and quality assessment of packaged spinach treated with a novel ozone-generation system. LWT – Food Science and Technology. 2009;**42**:1047-1053. DOI: 10.1016/j.lwt.2009.02.011

[79] Zhang L, Lu Z, Yu Z, Gao X. Preservation of fresh-cut celery by treatment of ozonated water. Food Control. 2005;**16**:279-283. DOI: 10.1016/j.foodcont.2004.03.007

[80] Rico D, Martín-Diana AB, Frías JM, Henehan GTM, Barry-Ryan C. Effect of ozone and calcium lactate treatments on browning and texture properties of fresh-cut lettuce. Journal of the Science of Food and Agriculture. 2006;**86**:2179-2188. DOI: 10.1002/jsfa.2594

[81] Gutiérrez DR, Chaves AR, Rodríguez SDC. Use of UV-C and gaseous ozone as sanitizing agents for keeping the quality of fresh-cut rocket (Eruca sativa mill). Journal of Food Processing and Preservation. 2017;**41e**:1-13. DOI: 10.1111/jfpp.12968

[82] Aguayo E, Escalona VH, Artés F. Effect of cyclic exposure to ozone gas on physicochemical, sensorial and microbial quality of whole and sliced tomatoes. Postharvest Biology and Technology. 2006;**39**:169-177. DOI: 10.1016/j.postharvbio.2005.11.005

[83] Chen J, Hu Y, Wang J, Hu H, Cui H. Combined effect of ozone treatment and modified atmosphere packaging on antioxidant defense system of fresh-cut green peppers. Journal of Food Processing and Preservation. 2016;**40**:1145-1150. DOI: 10.1111/jfpp.12695

[84] Silveira AC, Aguayo E, Artés F. Emerging sanitizers and clean room packaging for improving the microbial quality of fresh-cut 'Galia' melon. Food Control. 2010;**21**:863-871. DOI: 10.1016/j.foodcont.2009.11.017

[85] Baur S, Klaiber R, Hammes WP, Carle R. Sensory and microbiological quality of shredded, packaged iceberg lettuce as affected by pre-washing procedures with chlorinated and ozonated water. Innovative Food Science & Emerging Technologies. 2004;**5**:45-55. DOI: 10.1016/j.ifset.2003.10.002

[86] Burt S. Essential oils: Their antibacterial properties and potential applications in foods-A review. International Journal of Food Microbiology. 2004;**94**:223-253. DOI: 10.1016/j. ijfoodmicro.2004.03.022

[87] Hirasa K, Takemasa M, Antimicrobial and antioxidant properties of spices. In: Hirasa K, Takemasa M, editors. Spice Science and Technology. New York (USA): Marcel Dekker Inc.; 1998. pp. 163-200

[88] Rota C, Carraminana JJ, Burillo J, Herrera A. *In vitro* antimicrobial activity of essential oils from aromatic plants against selected foodborne pathogens. Journal of Food Protection. 2004;**67**:1252-1256. DOI: 10.4315/0362-028X-67.6.1252

[89] Nychas GJE, Skandamis PN, Tassou CC, Antimicrobials from herbs and spices. In: Roller S, editor. Natural Antimicrobials for the Minimal Processing of Foods. Boca Raton FL (USA): CRC Press; 2000

[90] Bagamboula CF, Uyttendaele M, Debevere J. Inhibitory effect of thyme and basil essential oils, carvacrol, thymol, estragol, linalool and p-cymene towards *Shigella sonnei* and *S. flexneri*. Food Microbiology. 2004;**21**:33-42. DOI: 10.1016/S0740-0020(03)00046-7

[91] Onursal CE, Eren I, Güneyli A, Topcu T, Çalhan O, Bayindir D. Effect of Carvacrol on Microbial Activity and Storage Quality of Fresh-cut 'Braeburn' Apple. Leuven, Belgium: International Society for Horticultural Science (ISHS); 2014. pp. 215-221

[92] Roller S, Seedhar P. Carvacrol and cinnamic acid inhibit microbial growth in fresh-cut melon and kiwifruit at 4 and 8°C. Letters in Applied Microbiology. 2002;**35**:390-394. DOI: 10.1046/j.1472-765X.2002.01209.x

[93] Valero D, Valverde JM, Martínez-Romero D, Guillén F, Castillo S, Serrano M. The combination of modified atmosphere packaging with eugenol or thymol to maintain quality, safety and functional properties of table grapes. Postharvest Biology and Technology. 2006;**41**:317-327. DOI: 10.1016/j.postharvbio.2006.04.011

[94] Salvia-Trujillo L, Rojas-Graü A, Soliva-Fortuny R, Martín-Belloso O. Physicochemical characterization and antimicrobial activity of food-grade emulsions and nanoemulsions incorporating essential oils. Food Hydrocolloids. 2015;**43**:547-556. DOI: 10.1016/j.foodhyd.2014.07.012

[95] Martínez-Hernández GB, Amodio ML, Colelli G. Carvacrol-loaded chitosan nanoparticles maintain quality of fresh-cut carrots. Innovative Food Science & Emerging Technologies. 2017;**41**:56-63. DOI: 10.1016/j.ifset.2017.02.005

[96] Gao H, Zhou Y, Fang X, Mu H, Han Q, Chen H-J. Development and characterization of an antimicrobial packaging film coating containing AITC or carvacrol for preservation of fresh-cut vegetable. In: Book of Abstracts of the III International Conference on Fresh-Cut Produce: Maintaining Quality and Safety; 13-18 September 2015; Davis. Davis: UC Davis; 2015. p. 175

[97] Tiwari BK, Valdramidis VP, O' Donnell CP, Muthukumarappan K, Bourke P, Cullen PJ. Application of natural antimicrobials for food preservation. Journal of Agricultural and Food Chemistry. 2009;**57**:5987-6000. DOI: 10.1021/jf900668n

[98] Jang M, Hong E, Kim GH. Evaluation of antibacterial activity of 3-butenyl, 4-pentenyl, 2-phenylethyl, and benzyl isothiocyanate in *Brassica* vegetables. Journal of Food Science. 2010;**75**:412-416. DOI: 10.1111/j.1750-3841.2010.01725.x

[99] Sofrata A, Santangelo EM, Azeem M, Borg-Karlson A-K, Gustafsson A, Pütsep K. Benzyl isothiocyanate, a major component from the roots of salvadora persica is highly

active against gram-negative bacteria. PLoS One. 2011;**6**:e23045. DOI: 10.1371/journal. pone.0023045

[100] EFSA. EFSA panel on food additives and nutrient sources added to food (ANS): Scientific opinion on the safety of allyl isothiocyanate for the proposed uses as a food additive. EFSA Journal. 2010;**8**:1943-1983

[101] Pablos C, Fernández A, Thackeray A, Marugán J. Effects of natural antimicrobials on prevention and reduction of bacterial cross-contamination during the washing of ready-to-eat fresh-cut lettuce. Food Science and Technology International. 2017;**23**:403-414. DOI: 10.1177/1082013217697851

[102] Galvez A, Abriouel H, Lopez RL, Ben Omar N. Bacteriocin-based strategies for food biopreservation. International Journal of Food Microbiology. 2007;**120**:51-70. DOI: 10.1016/j.ijfoodmicro.2007.06.001

[103] Bari ML, Ukuku DO, Kawasaki T, Inatsu Y, Isshiki K, Kawamoto S. Combined efficacy of nisin and pediocin with sodium lactate, citric acid, phytic acid, and potassium sorbate and EDTA in reducing the *Listeria monocytogenes* population of inoculated fresh-cut produce. Journal of Food Protection. 2005;**68**:1381-1387. DOI: 10.4315/0362-028X-68.7.1381

[104] Hansen JN, Sandine WE. Nisin as a model food preservative. Critical Reviews in Food Science and Nutrition. 1994;**34**:69-93. DOI: 10.1080/10408399409527650

[105] Silveira AC, Conesa A, Aguayo E, Artés F. Alternative sanitizers to chlorine for use on fresh-cut "Galia" (*Cucumis melo* var. catalupensis) melon. Journal of Food Science. 2008;**73**:M405–M411. DOI: 10.1111/j.1750-3841.2008.00939.x

[106] Allende A, Martinez B, Selma V, Gil MI, Suarez JE, Rodriguez A. Growth and bacteriocin production by lactic acid bacteria in vegetable broth and their effectiveness at reducing Listeria monocytogenes in vitro and in fresh-cut lettuce. Food Microbiology. 2007;**24**:759-766. DOI: 10.1016/j.fm.2007.03.002

[107] Sanchís E, González S, Ghidelli C, Sheth CC, Mateos M, Palou L, Pérez-Gago MB. Browning inhibition and microbial control in fresh-cut persimmon (*Diospyros kaki* Thunb. cv. Rojo Brillante) by apple pectin-based edible coatings. Postharvest Biology and Technology. 2016;**112**:186-193. DOI: 10.1016/j.postharvbio.2015.09.024

[108] Lin H, Lin Y, Hung Y-C, Chen Y, Fan M. Effects of nisin treatment on microbial growth and quality of fresh-cut Chinese yam during storage. In: Book of Abstracts of the III International Conference on Fresh-Cut Produce: Maintaining Quality and Safety; 13-18 September 2015; Davis. Davis: UC Davis; 2015. p. 145

[109] Randazzo CL, Pitino I, Scifò GO, Caggia C. Biopreservation of minimally processed iceberg lettuces using a bacteriocin produced by *Lactococcus lactis* wild strain. Food Control. 2009;**20**:756-763. DOI: 10.1016/j.foodcont.2008.09.020

[110] Alegre I, Viñas I, Usall J, Anguera M, Figge MJ, Abadías M. An *Enterobacteriaceae* species isolated from apples controls foodborne pathogens on fresh-cut apples and peaches. Postharvest Biology and Technology. 2012;**74**:118-124. DOI: 10.1016/j. postharvbio.2012.07.004

[111] Trias R, Bañeras L, Badosa E, Montesinos E. Bioprotection of Golden Delicious apples and Iceberg lettuce against foodborne bacterial pathogens by lactic acid bacteria. International Journal of Food Microbiology. 2008;**123**:50-60. DOI: 10.1016/j.ijfoodmicro.2007.11.065

[112] Fan Y, Xu Y, Wang D, Zhang L, Sun J, Sun L, Zhang B. Effect of alginate coating combined with yeast antagonist on strawberry (Fragaria × ananassa) preservation quality. Postharvest Biology and Technology. 2009;**53**:84-90. DOI: 10.1016/j.postharvbio.2009.03.002

[113] Mari M, Martini C, Spadoni A, Rouissi W, Bertolini P. Biocontrol of apple postharvest decay by *Aureobasidium pullulans*. Postharvest Biology and Technology. 2012;**73**:56-62. DOI: 10.1016/j.postharvbio.2012.05.014

[114] Alegre I, Viñas I, Usall J, Teixidó N, Figge MJ, Abadías M. Control of foodborne pathogens on fresh-cut fruit by a novel strain of *Pseudomonas graminis*. Food Microbiology. 2013;**34**:390-399. DOI: 10.1016/j.fm.2013.01.013

[115] Abadías M, Altisent R, Usall J, Torres R, Oliveira M, Viñas I. Biopreservation of fresh-cut melon using the strain *Pseudomonas graminis* CPA-7. Postharvest Biology and Technology. 2014;**96**:69-77. DOI: 10.1016/j.postharvbio.2014.05.010

[116] Plaza L, Altisent R, Alegre I, Viñas I, Abadías M. Changes in the quality and antioxidant properties of fresh-cut melon treated with the biopreservative culture *Pseudomonas graminis* CPA-7 during refrigerated storage. Postharvest Biology and Technology. 2016;**111**:25-30. DOI: 10.1016/j.postharvbio.2015.07.023

[117] Simões M, Simões LC, Vieira MJ. A review of current and emergent biofilm control strategies. LWT – Food Science and Technology. 2010;**43**:573-583. DOI: 10.1016/j.lwt.2009.12.008

[118] Leverentz B, Conway WS, Camp MJ, Janisiewicz WJ, Abuladze T, Yang M, Saftner R, Sulakvelidze A. Biocontrol of *Listeria monocytogenes* on fresh-cut produce by treatment with lytic bacteriophages and a bacteriocin. Applied and Environmental Microbiology. 2003;**69**:4519-4526. DOI: 10.1128/aem.69.8.4519-4526.2003

[119] Spricigo DA, Bardina C, Cortes P, Llagostera M. Use of a bacteriophage cocktail to control *Salmonella* in food and the food industry. International Journal of Food Microbiology. 2013;**165**:169-174. DOI: 10.1016/j.ijfoodmicro.2013.05.009

[120] Vonasek E, Choi A, Sanchez J, Nitin N. Incorporating bacteriophages into edible dip coatings to control food pathogens on fresh produce. In: Book of Abstracts of the III International Conference on Fresh-Cut Produce: Maintaining Quality and Safety; 13-18 September 2015; Davis. Davis: UC Davis; 2015. p. 115

[121] Food and Drug Administration (FDA). Title 21: Food and Drugs. Section. Part 101: Food Labelling [Internet]. 2016. Available from: https://www.accessdata.fda.gov/scripts/cdrh/cfdocs/cfcfr/CFRSearch.cfm?CFRPart=101&showFR=1&subpartNode=21:2.0.1.1.2.6 [Accessed: 2017-07-10]

[122] Peña-Estévez ME, Gómez PA, Artés F, Aguayo E, Martínez-Hernández GB, Otón M, Galindo A, Artés-Hernández F. Quality changes of fresh-cut pomegranate arils during

shelf life as affected by deficit irrigation and postharvest vapour treatments. Journal of the Science of Food and Agriculture. 2015;**95**:2325-2336. DOI: 10.1002/jsfa.6954

[123] Maghoumi M, Mostofi Y, Zamani Z, Talaie A, Boojar M, Gómez PA. Influence of hot-air treatment, superatmospheric O_2 and elevated CO_2 on bioactive compounds and storage properties of fresh-cut pomegranate arils. International Journal of Food Science & Technology. 2014;**49**:153-159. DOI: 10.1111/ijfs.12290

[124] Beirão-da-Costa S, Empis J, Moldão-Martins M. Fresh-cut kiwifruit structure and firmness as affected by heat pre-treatments and post-cut calcium dips. Food and Bioprocess Technology. 2014;**7**:1128-1136. DOI: 10.1007/s11947-013-1151-3

[125] Martínez-Hernández GB, Gómez P, Orihuel-Iranzo B, Bretó J, Artés-Hernández F, Artés F. Innovative and sustainable postharvest treatments to control physiological disorders and decay in lemon fruit during long transport and commercialization. In: VIII International Postharvest Symposium, Cartagena. 2016

[126] Obando-Ulloa JM, Jiménez V, Machuca-Vargas A, Beaulieu JC, Infante R, Escalona-Contreras VH. Effect of hot water dips on the quality of fresh-cut Ryan Sun peaches. Idesia (Arica). 2015;**33**:13-26. DOI: 10.4067/S0718-34292015000100002

[127] Aguayo E, Requejo-Jackman C, Stanley R, Woolf A. Hot water treatment in combination with calcium ascorbate dips increases bioactive compounds and helps to maintain fresh-cut apple quality. Postharvest Biology and Technology. 2015;**110**:158-165. DOI: 10.1016/j.postharvbio.2015.07.001

[128] Wang Q, Nie X, Cantwell M. Hot water and ethanol treatments can effectively inhibit the discoloration of fresh-cut sunchoke (*Helianthus tuberosus* L.) tubers. Postharvest Biology and Technology. 2014;**94**:49-57. DOI: 10.1016/j.postharvbio.2014.03.003

[129] Murata M, Tanaka E, Minoura E, Homma S. Quality of cut lettuce treated by heat shock: Prevention of enzymatic browning, repression of phenylalanine ammonia–lyase activity, and improvement on sensory evaluation during. Bioscience, Biotechnology, and Biochemistry. 2004;**68**:501-507. DOI: 10.1271/bbb.68.501

[130] Hägele F, Baur S, Menegat A, Gerhards R, Carle R, Schweiggert RM. Chlorophyll fluorescence imaging for monitoring the effects of minimal processing and warm water treatments on physiological properties and quality attributes of fresh-cut salads. Food and Bioprocess Technology. 2016;**9**:650-663. DOI: 10.1007/s11947-015-1661-2

[131] Koukounaras A, Siomos AS, Sfakiotakis E. Impact of heat treatment on ethylene production and yellowing of modified atmosphere packaged rocket leaves. Postharvest Biology and Technology. 2009;**54**:172-176. DOI: 10.1016/j.postharvbio.2009.07.002

[132] Gómez F, Fernández L, Gergoff G, Guiamet JJ, Chaves A, Bartoli CG. Heat shock increases mitochondrial H_2O_2 production and extends postharvest life of spinach leaves. Postharvest Biology and Technology. 2008;**49**:229-234. DOI: 10.1016/j.postharvbio.2008.02.012

[133] Glowacz M, Mogren LM, Reade JPH, Cobb AH, Monaghan JM. Can hot water treatments enhance or maintain postharvest quality of spinach leaves? Postharvest Biology and Technology. 2013;**81**:23-28. DOI: 10.1016/j.postharvbio.2013.02.004

[134] Loaiza-Velarde JG, Mangrich ME, Campos-Vargas R, Saltveit ME. Heat shock reduces browning of fresh-cut celery petioles. Postharvest Biology and Technology. 2003;**27**:305-311. DOI: 10.1016/S0925-5214(02)00118-7

[135] Barbagallo RN, Chisari M, Caputa G. Effects of calcium citrate and ascorbate as inhibitors of browning and softening in minimally processed 'Birgah' eggplants. Postharvest Biology and Technology. 2012;**73**:107-114. DOI: 10.1016/j.postharvbio.2012.06.006

[136] Siddiq M, Roidoung S, Sogi DS, Dolan KD. Total phenolics, antioxidant properties and quality of fresh-cut onions (*Allium cepa* L.) treated with mild-heat. Food Chemistry. 2013;**136**:803-806. DOI: 10.1016/j.foodchem.2012.09.023

[137] Maghoumi M, Gómez PA, Mostofi Y, Zamani Z, Artés-Hernández F, Artés F. Combined effect of heat treatment, UV-C and superatmospheric oxygen packing on phenolics and browning related enzymes of fresh-cut pomegranate arils. LWT – Food Science and Technology. 2013;**54**:389-396. DOI: 10.1016/j.lwt.2013.06.006

[138] Peña-Estévez ME, Gómez PA, Artés F, Aguayo E, Martínez-Hernández GB, Galindo A, Torecillas A, Artés-Hernández F. Changes in bioactive compounds and oxidative enzymes of fresh-cut pomegranate arils during storage as affected by deficit irrigation and postharvest vapor heat treatments. Food Science and Technology International. 2016;**22**:665-676. DOI: 10.1177/1082013216635323

[139] Maghoumi M, Gomez PA, Artes-Hernandez F, Mostofi Y, Zamani Z, Artes F. Hot water, UV-C and superatmospheric oxygen packaging as hurdle techniques for maintaining overall quality of fresh-cut pomegranate arils. Journal of the Science of Food and Agriculture. 2013;**93**:1162-1168. DOI: 10.1002/jsfa.5868

[140] Martínez-Hernández GB, Amodio ML, Colelli G. Potential use of microwave treatment on fresh-cut carrots: physical, chemical and microbiological aspects. Journal of the Science of Food and Agriculture. 2016;**96**:2063-2072. DOI: 10.1002/jsfa.7319

[141] Gray NF. Ultraviolet disinfectio. In: Percival SL, Chalmers RM, Embrey M, Hunter P, Sellwood JPW-J, editors. Microbiology of Waterborne Diseases. London: Academic Press; 2014. pp. 617-630.

[142] Bintsis T, Litopoulou-Tzanetaki E, Robinson RK. Existing and potential applications of ultraviolet light in the food industry – A critical review. Journal of the Science of Food and Agriculture. 2000;**80**:637-645. DOI: 10.1002/(SICI)1097-0010(20000501)80:6<637::AID-JSFA603>3.0.CO;2-1

[143] Yaun BR, Sumner SS, Eifert JD, Marcy JE. Inhibition of pathogens on fresh produce by ultraviolet energy. International Journal of Food Microbiology. 2004;**90**:1-8. DOI: 10.1016/S0168-1605(03)00158-2

[144] Sharma G. Ultraviolet light. In: Robinson RK, Batt C, Patel P, editors. Encyclopedia of Food Microbiology-3. London: Academic Press; 1999. pp. 2208-2214

[145] Ben Said M, Masahiro O, Hassen A. Detection of viable but non cultivable Escherichia coli after UV irradiation using a lytic Qβ phage. Annals of Microbiology. 2010;**60**:121-127. DOI: 10.1007/s13213-010-0017-4

[146] Formica-Oliveira AC, Martínez-Hernández GB, Aguayo E, Gómez PA, Artés F, Artés-Hernández F. UV-C and hyperoxia abiotic stresses to improve healthiness of carrots: Study of combined effects. Journal of Food Science and Technology. 2016;**53**:1-12. DOI: 10.1007/s13197-016-2321-x

[147] Hadjok C, Mittal GS, Warriner K. Inactivation of human pathogens and spoilage bacteria on the surface and internalized within fresh produce by using a combination of ultraviolet light and hydrogen peroxide. Journal of Applied Microbiology. 2008;**104**:1014-1024. DOI: 10.1111/j.1365-2672.2007.03624.x

[148] Kim Y-H, Jeong S-G, Back K-H, Park K-H, Chung M-S, Kang D-H. Effect of various conditions on inactivation of Escherichia coli O157:H7, Salmonella Typhimurium, and Listeria monocytogenes in fresh-cut lettuce using ultraviolet radiation. International Journal of Food Microbiology. 2013;**166**:349-355. DOI: 10.1016/j.ijfoodmicro.2013.08.010

[149] Martínez-Hernández GB, Huertas J-P, Navarro-Rico J, Gómez PA, Artés F, Palop A, Artés-Hernández F. Inactivation kinetics of foodborne pathogens by UV-C radiation and its subsequent growth in fresh-cut kailan-hybrid broccoli. Food Microbiology. 2015;**46**:263-271. DOI: 10.1016/j.fm.2014.08.008

[150] Artés-Hernández F, Escalona VH, Robles PA, Martínez-Hernández GB, Artés F. Effect of UV-C radiation on quality of minimally processed spinach leaves. Journal of the Science of Food and Agriculture. 2009;**89**:414-421. DOI: 10.1002/jsfa.3460

[151] European Union (EU). Commission regulation (EC) No 2073/2005 of 15 November 2005 on microbiological criteria for foodstuffs. Official Journal of the European Union. 2005;L338:1-26

[152] Escalona VH, Aguayo E, Martínez-Hernández GB, Artés F. UV-C doses to reduce pathogen and spoilage bacterial growth in vitro and in baby spinach. Postharvest Biology and Technology. 2010;**56**:223-231. DOI: 10.1016/j.postharvbio.2010.01.008

[153] Erkan M, Wang CY, Krizek DT. UV-C irradiation reduces microbial populations and deterioration in Cucurbita pepo fruit tissue. Environmental and Experimental Botany. 2001;**45**:1-9. DOI: S0098-8472(00)00073-3

[154] Martínez-Hernández GB, Gómez PA, Pradas I, Artés F, Artés-Hernández F. Moderate UV-C pretreatment as a quality enhancement tool in fresh-cut Bimi® broccoli. Postharvest Biology and Technology. 2011;**62**:327-337. DOI: 10.1016/j.postharvbio.2011.06.015

[155] Robles P, De Campos A, Artés-Hernández F, Gómez P, Calderón A, Ferrer M, Artés F. Combined effect of UV–C radiation and controlled atmosphere storage to preserve tomato quality. In: V Congreso Iberoamericano de Tecnología Postcosecha y Agroexportaciones; Cartagena (Spain). 2007

[156] Jemni M, Gómez PA, Souza M, Chaira N, Ferchichi A, Otón M, Artés F. Combined effect of UV-C, ozone and electrolyzed water for keeping overall quality of date palm. LWT – Food Science and Technology. 2014;59:649-655. DOI: 10.1016/j.lwt.2014.07.016

[157] Pinheiro JC, Alegria CSM, Abreu MMMN, Gonçalves EM, Silva CLM. Evaluation of alternative preservation treatments (water heat treatment, ultrasounds, thermosonication and UV-C radiation) to improve safety and quality of whole tomato. Food and Bioprocess Technology. 2016;9:924-935. DOI: 10.1007/s11947-016-1679-0

[158] Marquenie D, Michiels CW, Geeraerd AH, Schenk A, Soontjens C, Van Impe JF, Nicolaï BM. Using survival analysis to investigate the effect of UV-C and heat treatment on storage rot of strawberry and sweet cherry. International Journal of Food Microbiology. 2002;73:187-196. DOI: 10.1016/S0168-1605(01)00648-1

[159] Artés-Hernández F, Robles PA, Gómez PA, Tomás-Callejas A, Artés F. Low UV-C illumination for keeping overall quality of fresh-cut watermelon. Postharvest Biology and Technology. 2010;55:114-120. DOI: 10.1016/j.postharvbio.2009.09.002

[160] Rocha ABO, Honório SL, Messias CL, Otón M, Gómez PA. Effect of UV-C radiation and fluorescent light to control postharvest soft rot in potato seed tubers. Scientia Horticulturae. 2015;181:174-181. DOI: 10.1016/j.scienta.2014.10.045

[161] Allende A, Artés F. UV-C radiation as a novel technique for keeping quality of fresh processed 'Lollo Rosso' lettuce. Food Research International. 2003;36:739-746. DOI: 10.1016/S0963-9969(03)00054-1

[162] Fan X, Toivonen PM, Rajkowski KT, Sokorai KJ. Warm water treatment in combination with modified atmosphere packaging reduces undesirable effects of irradiation on the quality of fresh-cut iceberg lettuce. Journal of Agricultural and Food Chemistry. 2003;51:1231-1236. DOI: 10.1021/jf020600c

[163] González-Aguilar GA, Zavaleta-Gatica R, Tiznado-Hernández ME. Improving postharvest quality of mango 'Haden' by UV-C treatment. Postharvest Biology and Technology. 2007;45:108-116. DOI: 10.1016/j.postharvbio.2007.01.012

[164] Alothman M, Bhat R, Karim AA. Effects of radiation processing on phytochemicals and antioxidants in plant produce. Trends in Food Science & Technology. 2009;20:201-212. DOI: 10.1016/j.tifs.2009.02.003

[165] Rawson A, Patras A, Tiwari BK, Noci F, Koutchma T, Brunton N. Effect of thermal and non thermal processing technologies on the bioactive content of exotic fruits and their products: Review of recent advances. Food Research International. 2011;44:1875-1887. DOI: 10.1016/j.foodres.2011.02.053

[166] Scattino C, Castagna A, Neugart S, Chan HM, Schreiner M, Crisosto CH, Tonutti P, Ranieri A. Post-harvest UV-B irradiation induces changes of phenol contents and corresponding biosynthetic gene expression in peaches and nectarines. Food Chemistry. 2014;163:51-60. DOI: 10.1016/j.foodchem.2014.04.077

[167] Castagna A, Dall'Asta C, Chiavaro E, Galaverna G, Ranieri A. Effect of post-harvest UV-B irradiation on polyphenol profile and antioxidant activity in flesh and peel of tomato fruits. Food Bioprocess Technology. 2014;7:2241-2250. DOI: 10.1007/s11947-013-1214-5

[168] Du WX, Avena-Bustillos RJ, Breksa 3rd AP, McHugh TH. Effect of UV-B light and different cutting styles on antioxidant enhancement of commercial fresh-cut carrot products. Food Chemistry. 2012;134:1862-1869. DOI: 10.1016/j.foodchem.2012.03.097

[169] Formica-Oliveira AC, Martínez-Hernández GB, Díaz-López V, Artés F, Artés-Hernández F. Effects of UV-B and UV-C combination on phenolic compounds biosynthesis in fresh-cut carrots. Postharvest Biology and Technology. 2017;127:99-104. DOI: 10.1016/j.postharvbio.2016.12.010

[170] Kataoka I, Beppu K, Sugiyama A, Taira S. Enhancement of cooration of Satohnishiki sweet cherry fruit by postharvest irradiation with ultraviolet rays. Environment Control in Biology. 1996;34:313-319. DOI: 10.2525/ecb1963.34.313

[171] Woltering EJ, Witkowska IM, Schouten R, Harbinson J. Low intensity postharvest lighting improves quality and shelf life of fresh-cut lettuce. In: Book of Abstracts of the III International Conference on Fresh-Cut Produce: Maintaining Quality and Safety; 13-18 September 2015; Davis. Davis: UC Davis; 2015. p. 115

[172] Oms-Oliu G, Martín-Belloso O, Soliva-Fortuny R. Pulsed light treatments for food preservation. A review. Food and Bioprocess Technology. 2008;3:13. DOI: 10.1007/s11947-008-0147-x

[173] Gómez-López VM, Ragaert P, Debevere J, Devlieghere F. Pulsed light for food decontamination: A review. Trends in Food Science and Technology. 2007;18:464-473. DOI: 10.1016/j.tifs.2007.03.010

[174] Ramos-Villarroel AY, Martín-Belloso O, Soliva-Fortuny R. Bacterial inactivation and quality changes in fresh-cut avocado treated with intense light pulses. European Food Research and Technology. 2011;233:395-402. DOI: 10.1007/s00217-011-1533-6

[175] Ramos-Villarroel AY, Aron-Maftei N, Martín-Belloso O, Soliva-Fortuny R. Influence of spectral distribution on bacterial inactivation and quality changes of fresh-cut watermelon treated with intense light pulses. Postharvest Biology and Technology. 2012;69:32-39. DOI: 10.1016/j.postharvbio.2012.03.002

[176] Ramos-Villarroel AY, Aron-Maftei N, Martín-Belloso O, Soliva-Fortuny R. The role of pulsed light spectral distribution in the inactivation of Escherichia coli and Listeria innocua on fresh-cut mushrooms. Food Control. 2012;24:206-213. DOI: 10.1016/j.foodcont.2011.09.029

[177] Oms-Oliu G, Aguiló-Aguayo I, Martín-Belloso O, Soliva-Fortuny R. Effects of pulsed light treatments on quality and antioxidant properties of fresh-cut mushrooms (Agaricus bisporus). Postharvest Biology and Technology. 2010;56:216-222. DOI: 10.1016/j.postharvbio.2009.12.011

[178] Izquier A, Gómez-López VM. Modeling the pulsed light inactivation of microorganisms naturally occurring on vegetable substrates. Food Microbiology. 2011;28:1170-1174. DOI: 10.1016/j.fm.2011.03.010

[179] Agüero MV, Jagus RJ, Martín-Belloso O, Soliva-Fortuny R. Surface decontamination of spinach by intense pulsed light treatments: Impact on quality attributes. Postharvest Biology and Technology. 2016;**121**:118-125. DOI: 10.1016/j.postharvbio.2016.07.018

[180] Gómez-López VM, Devlieghere F, Bonduelle V, Debevere J. Intense light pulses decontamination of minimally processed vegetables and their shelf-life. International Journal of Food Microbiology. 2005;**103**:79-89. DOI: 10.1016/j.ijfoodmicro.2004.11.028

[181] Lee E, Lee H, Jung W, Park S, Yang D, Lee K. Influences of humic acids and photoreactivation on the disinfection of *Escherichia coli* by a high-power pulsed UV irradiation. Korean Journal of Chemical Engineering. 2009;**26**:1301-1307. DOI: 10.1007/s11814-009-0208-5

[182] Lasagabaster A, de Maranon IM. Survival and growth of *Listeria innocua* treated by pulsed light technology: Impact of post-treatment temperature and illumination conditions. Food Microbiology. 2014;**41**:76-81. DOI: 10.1016/j.fm.2014.02.001

[183] Maclean M, Murdoch LE, Lani MN, MacGregor SJ, Anderson JG, Woolsey GA. Photoinactivation and photoreactivation responses by bacterial pathogens after exposure to pulsed UV-light. In: Proceedings of the 2008 IEEE International Power Modulators and High-Voltage Conference; 27-31 May 2008; Las Vegas. Boston: IEE; 2008. pp. 326-329

[184] Gomez-Lopez VM, Devlieghere F, Bonduelle V, Debevere J. Factors affecting the inactivation of micro-organisms by intense light pulses. Journal of Applied Microbiology. 2005;**99**:460-470. DOI: 10.1111/j.1365-2672.2005.02641.x

[185] Asavasanti S, Ersus S, Ristenpart W, Stroeve P, Barrett DM. Critical electric field strengths of onion tissues treated by pulsed electric fields. Journal of Food Science. 2010;**75**:E433–E443. DOI: 10.1111/j.1750-3841.2010.01768.x

[186] Lebovka NI, Praporscic I, Vorobiev E. Combined treatment of apples by pulsed electric fields and by heating at moderate temperature. Journal of Food Engineering. 2004;**65**:211-217. DOI: 10.1016/j.jfoodeng.2004.01.017

[187] Jin TZ, Yu Y, Gurtler JB. Effects of pulsed electric field processing on microbial survival, quality change and nutritional characteristics of blueberries. LWT – Food Science and Technology. 2017;**77**:517-524. DOI: 10.1016/j.lwt.2016.12.009

[188] Scholtz V, Pazlarova J, Souskova H, Khun J, Julak J. Nonthermal plasma-A tool for decontamination and disinfection. Biotechnology Advances. 2015;**33**:1108-1119. DOI: 10.1016/j.biotechadv.2015.01.002

[189] Dey A, Rasane P, Choudhury A, Singh J, Maisnam D, Rasane P. Cold plasma processing: A review. Journal of Chemical and Pharmaceutical Research. 2016;**9**:2980-2984

[190] Thirumdas R, Sarangapani C, Annapure US. Cold plasma: A novel non-thermal technology for food processing. Food Biophysics. 2015;**10**:1-11. DOI: 10.1007/s11483-014-9382-z

[191] Ziuzina D, Han L, Cullen PJ, Bourke P. Cold plasma inactivation of internalised bacteria and biofilms for *Salmonella enterica* serovar Typhimurium, *Listeria monocytogenes* and *Escherichia coli*. International Journal of Food Microbiology. 2015;**210**:53-61. DOI: 10.1016/j.ijfoodmicro.2015.05.019

[192] Li X, Farid M. A review on recent development in non-conventional food sterilization technologies. Journal of Food Engineering. 2016;**182**:33-45. DOI: 10.1016/j.jfoodeng.2016.02.026

[193] Ehlbeck J, Schnabel U, Polak M, Winter J, Woedtke T, Brandenburg R, Dem Hagen T, Weltmann K-D. Low temperature atmospheric pressure plasma sources for microbial decontamination. Journal of Physics D-Applied Physics. 2011;**44**:1-18. DOI: 10.1088/0022-3727/44/1/013002

[194] Pignata C, D'Angelo D, Fea E, Gilli G. A review on microbiological decontamination of fresh produce with nonthermal plasma. Journal of Applied Microbiology. 2017. DOI: 10.1111/jam.13412

[195] Tappi S, Berardinelli A, Ragni L, Dalla Rosa M, Guarnieri A, Rocculi P. Atmospheric gas plasma treatment of fresh-cut apples. Innovative Food Science & Emerging Technologies. 2014;**21**:114-122. DOI: 10.1016/j.ifset.2013.09.012

[196] Bußler S, Ehlbeck J, Schlüter OK. Pre-drying treatment of plant related tissues using plasma processed air: Impact on enzyme activity and quality attributes of cut apple and potato. Innovative Food Science & Emerging Technologies. 2017;**40**:78-86. DOI: 10.1016/j.ifset.2016.05.007

[197] Liao X, Liu D, Xiang Q, Ahn J, Chen S, Ye X, Ding T. Inactivation mechanisms of nonthermal plasma on microbes: A review. Food Control. 2017;**75**:83-91. DOI: 10.1016/j.foodcont.2016.12.021

[198] Otto C, Zahn S, Rost F, Zahn P, Jaros D, Rohm H. Physical methods for cleaning and disinfection of surfaces. Food Engineering Reviews. 2011;**3**:171-188. DOI: 10.1007/s12393-011-9038-4

[199] São José JFBd, Andrade NJd, Ramos AM, Vanetti MCD, Stringheta PC, Chaves JBP. Decontamination by ultrasound application in fresh fruits and vegetables. Food Control. 2014;**45**:36-50. DOI: 10.1016/j.foodcont.2014.04.015

[200] Health Protection Agency (HPA). Health Effects of Exposure to Ultrasound and Infrasound, Report of the independent advisory group on non-ionising radiation. London: HPA; 2010. 180. DOI: 978-0-85951-662-4

[201] Paniwnyk L. Application of ultrasound. In: Sun D-W, editor. Emerging Technologies for Food Processing. San Diego CA (USA): Academic Press; 2014. pp. 271-291

[202] Seymour IJ, Burfoot D, Smith RL, Cox LA, Lockwood A. Ultrasound decontamination of minimally processed fruits and vegetables. International Journal of Food Science & Technology. 2002;**37**:547-557. DOI: 10.1046/j.1365-2621.2002.00613.x

[203] do Rosario DK, da Silva Mutz Y, Peixoto JM, Oliveira SB, de Carvalho RV, Carneiro JC, de Sao Jose JF, Bernardes PC. Ultrasound improves chemical reduction of natural contaminant microbiota and *Salmonella enterica* subsp. enterica on strawberries. International Journal of Food Microbiology. 2017;**241**:23-29. DOI: 10.1016/j.ijfoodmicro.2016.10.009

[204] Birmpa A, Sfika V, Vantarakis A. Ultraviolet light and Ultrasound as non-thermal treatments for the inactivation of microorganisms in fresh ready-to-eat foods. International Journal of Food Microbiology. 2013;167:96-102. DOI: 10.1016/j.ijfoodmicro.2013.06.005

[205] Kim HJ, Feng H, Kushad MM, Fan X. Effects of ultrasound, irradiation, and acidic electrolyzed water on germination of alfalfa and broccoli seeds and *Escherichia coli* O157:H7. Journal of Food Science. 2006;71:168-173. DOI: 10.1111/j.1750-3841.2006.00064.x

[206] Jung S, Samson CT, Lamballerie M. High hydrostatic pressure food processing. In: Proctor A, editor. Alternatives to Conventional Food Processing. London, UK: Royal Society of Chemistry Publishing; 2011. pp. 254-306.

[207] Rendueles E, Omer MK, Alvseike O, Alonso-Calleja C, Capita R, Prieto M. Microbiological food safety assessment of high hydrostatic pressure processing: A review. LWT – Food Science and Technology. 2011;44:1251-1260. DOI: 10.1016/j.lwt.2010.11.001

[208] Préstamo G, Arroyo G. High hydrostatic pressure effects on vegetable structure. Journal of Food Science. 1998;63:878-881. DOI: 10.1111/j.1365-2621.1998.tb17918.x

[209] Oey I, Lille M, Van Loey A, Hendrickx M. Effect of high-pressure processing on colour, texture and flavour of fruit- and vegetable-based food products: A review. Trends in Food Science and Technology. 2008;19:320-328. DOI: 10.1016/j.tifs.2008.04.001

[210] Vázquez-Gutierrez JL, Quiles A, Vonasek E, Jernstedt JA, Hernando I, Nitin N, Barrett DM. High hydrostatic pressure as a method to preserve fresh-cut Hachiya persimmons: A structural approach. Food Science and Technology International. 2016;22:688-698. DOI: 10.1177/1082013216642049

[211] Denoya GI, Polenta GA, Apóstolo NM, Budde CO, Sancho AM, Vaudagna SR. Optimization of high hydrostatic pressure processing for the preservation of minimally processed peach pieces. Innovative Food Science & Emerging Technologies. 2016;33:84-93. DOI: 10.1016/j.ifset.2015.11.014

[212] Spilimbergo S, Komes D, Vojvodic A, Levaj B, Ferrentino G. High pressure carbon dioxide pasteurization of fresh-cut carrot. The Journal of Supercritical Fluids. 2013;79:92-100. DOI: 10.1016/j.supflu.2012.12.002

[213] Ferrentino G, Komes D, Spilimbergo S. High-power ultrasound assisted high-pressure carbon dioxide pasteurization of fresh-cut coconut: A microbial and physicochemical study. Food and Bioprocess Technology. 2015;8:2368-2382. DOI: 10.1007/s11947-015-1582-0

[214] Kader AA, Ben-Yehoshua S. Effects of superatmospheric oxygen levels on postharvest physiology and quality of fresh fruits and vegetables. Postharvest Biology and Technology. 2000;20:1-13. DOI: 10.1016/S0925-5214(00)00122-8

[215] Escalona VH, Geysen S, Verlinden BE, Nicolaï BM. Microbial quality and browning of fresh-cut butter lettuce under superatmospheric oxygen condition. European Journal of Horticultural Science. 2007;72:130-137

[216] Geysen S, Escalona VH, Verlinden BE, Aertsen A, Geeraerd AH, Michiels CW, Van Impe JF, Nicolaï BM. Validation of predictive growth models describing superatmospheric oxygen effects on *Pseudomonas fluorescens* and *Listeria innocua* on fresh-cut lettuce. International Journal of Food Microbiology. 2006;**111**:48-58. DOI: 10.1016/j.ijfoodmicro.2006.04.044

[217] Zheng Y, Yang Z, Chen X. Effect of high oxygen atmospheres on fruit decay and quality in Chinese bayberries, strawberries and blueberries. Food Control. 2008;**19**:470-474. DOI: 10.1016/j.foodcont.2007.05.011

[218] Allende A, Luo Y, McEvoy JL, Artés F, Wang CY. Microbial and quality changes in minimally processed baby spinach leaves stored under super atmospheric oxygen and modified atmosphere conditions. Postharvest Biology and Technology. 2004;**33**:51-59. DOI: 10.1016/j.postharvbio.2004.03.003

[219] Van der Steen C, Jacxsens L, Devlieghere F, Debevere J. Combining high oxygen atmospheres with low oxygen modified atmosphere packaging to improve the keeping quality of strawberries and raspberries. Postharvest Biology and Technology. 2002;**26**:49-58. DOI: 10.1016/S0925-5214(02)00005-4

[220] Heimdal H, Kühn BF, Poll L, Larsen LM. Biochemical changes and sensory quality of shredded and MA-packaged iceberg lettuce. Journal of Food Science. 1995;**60**:1265-1268. DOI: 10.1111/j.1365-2621.1995.tb04570.x

[221] Jacxsens L, Devlieghere F, Van der Steen C, Debevere J. Effect of high oxygen modified atmosphere packaging on microbial growth and sensorial qualities of fresh-cut produce. International Journal of Food Microbiology. 2001;**71**:197-210. DOI: 10.1016/S0168-1605(01)00616-X

[222] Oms-Oliu G, Soliva-Fortuny R, Martín-Belloso O. Modeling changes of headspace gas concentrations to describe the respiration of fresh-cut melon under low or superatmospheric oxygen atmospheres. Journal of Food Engineering. 2008;**85**:401-409. DOI: 10.1016/j.jfoodeng.2007.08.001

[223] Lu C, Toivonen PMA. Effect of 1 and 100 kPa O_2 atmospheric pretreatments of whole 'Spartan' apples on subsequent quality and shelf life of slices stored in modified atmosphere packages. Postharvest Biology and Technology. 2000;**18**:99-107. DOI: 10.1016/S0925-5214(99)00069-1

[224] Amanatidou A, Slump RA, Gorris LGM, Smid EJ. High oxygen and high carbon dioxide modified atmospheres for shelf-life extension of minimally processed carrots. Journal of Food Science. 2000;**65**:61-66. DOI: 10.1111/j.1365-2621.2000.tb15956.x

[225] Formica-Oliveira AC, Martínez-Hernández GB, Aguayo E, Gómez PA, Artés F, Artés-Hernández F. A functional smoothie from carrots with induced enhanced phenolic content. Food and Bioprocess Technology. 2017;**10**:491-502. DOI: 10.1007/s11947-016-1829-4

[226] Gorny J, Agar I. Are argon-enriched atmospheres beneficial? Perishables Handling Newsletter. 1998;**94**:7-8

[227] Jamie P, Saltveit ME. Postharvest changes in broccoli and lettuce during storage in argon, helium, and nitrogen atmospheres containing 2% oxygen. Postharvest Biology and Technology. 2002;**26**:113-116. DOI: 10.1016/S0925-5214(02)00006-6

[228] Zhang M, Zhan ZG, Wang SJ, Tang JM. Extending the shelf-life of asparagus spears with a compressed mix of argon and xenon gases. LWT – Food Science and Technology. 2008;**41**:686-691. DOI: 10.1016/j.lwt.2007.04.011

[229] Tomás-Callejas A, Boluda M, Robles PA, Artés F, Artés-Hernández F. Innovative active modified atmosphere packaging improves overall quality of fresh-cut red chard baby leaves. LWT – Food Science and Technology. 2011;**44**:1422-1428. DOI: 10.1016/j.lwt.2011.01.020

[230] Gouble B, Fath D, Soudain P. Nitrous oxide inhibition of ethylene production in ripening and senescing climacteric fruits. Postharvest Biology and Technology. 1995;**5**:311-321. DOI: 10.1016/0925-5214(94)00030-V

[231] Rodríguez-Hidalgo S, Artés-Hernández F, Gómez PA, Fernández JA, Artés F. Quality of fresh-cut baby spinach grown under a floating trays system as affected by nitrogen fertilisation and innovative packaging treatments. Journal of the Science of Food and Agriculture. 2010;**90**:1089-1097. DOI: 10.1002/jsfa.3926

[232] Ansah FA, Amodio ML, Colelli G. Evaluation of the impact of nitrous oxide use on quality and shelf life of packaged fresh-cut 'iceberg' lettuce and wild rocket. Chemical Engineering Transactions. 2015;**44**:319-324. DOI: 10.3303/CET1544054

[233] Silveira AC, Araneda C, Hinojosa A, Escalona VH. Effect of non-conventional modified atmosphere packaging on fresh cut watercress (*Nasturtium officinale* R. Br.) quality. Postharvest Biology and Technology. 2014;**92**:114-120. DOI: 10.1016/j.postharvbio.2013.12.012

[234] Inestroza-Lizardo C, Silveira AC, Escalona VH. Metabolic activity, microbial growth and sensory quality of arugula leaves (*Eruca vesicaria* Mill.) stored under non-conventional modified atmosphere packaging. Scientia Horticulturae. 2016;**209**:79-85. DOI: 10.1016/j.scienta.2016.06.007

8

Postharvest Handling of Indigenous and Underutilized Fruits in Trinidad and Tobago

Puran Bridgemohan and Wendy-Ann P. Isaac

Additional information is available at the end of the chapter

Abstract

This chapter briefly outlines the origin of some indigenous and underutilized fruit crops found throughout Trinidad and Tobago. It also examines the current situation, current practices, and maturity standards for postharvest handling of these commodities and examines the principle causes of postharvest losses and poor quality. Finally, the chapter includes some recommendations on the best postharvest practices for these indigenous and underutilized fruits, including field harvesting practices, storage and transportation, and cool storage.

Keywords: indigenous, underutilized fruits, postharvest handling

1. Introduction

The present topic is characterized by a large natural population of indigenous and underutilized fruits which despite growing in popularity are on the verge of being extinct as they are not cultivated in large acreages. Several of these crops were consumed as the main fruits by earlier inhabitants in the Caribbean. However, most of them have lost their popularity due to changes in the taste patterns in favor of temperate fruits such as grapes, apples, pears, peaches, and plums. With the growing demand for some of these commodities and possible export potential, there are considerable market development opportunities within the Caribbean for these fruits. The low production of these fruits and the limited information on the postharvest physiology and disease characteristics of these fruits have made large-scale production and packing/processing difficult. The magnitude of postharvest losses of many of these fruits can exceed 50% depending on the commodity as many of them are highly perishable due to high moisture content under tropical conditions [1].

Many of these fruits are harvested, handled, and sometimes processed using local indigenous knowledge at the household level. The main ripe fruits that fall in this category are the edible fresh fruits and include pommerac *(Syzygium malaccense)*, sapodilla *(Manilkara zapota)*, balata *(Manilkara bidentata)*, carambola *(Averrhoa carambola)*, guinep *(Melicoccus bijugatus)*, and cashew *(Anacardium occidentale)*. The fruits that are harvested mature but do require to be ripened and need to be cooked include breadfruit *(Artocarpus altilis)* and chataigne *(Artocarpus camansi)* and also the palm fruit peewah *(Bactris gasipaes)*. There are two edible fresh fruits that are made into frappes or sherbets and ice cream, sweet sop *(Annona squamosa)* and barbadine *(Passiflora quadrangularis)*. There is also a very popular root tuber in the Caribbean Latin American countries called topi tamboor leren *(Calathea allouia)*. **Table 1** shows some of the most common indigenous and underutilized fruits in Trinidad and Tobago, their utilization, harvest and postharvest practices.

Crop	Utilization	Harvesting and postharvest handling
Calathea allouia (Aubl.) Lindl. (topi tambo) Plate 1	• Eaten boiled as an appetizer • Tastes similar to water chestnut and can be used as a substitute • Retains its crispness when cooked • Can be used to make flours	• Harvested after 9 months by using fork to pull up tubers • Tubers can be stored at room temperatures for up to 3 months • Does not tolerate refrigeration • Spoilage in plastic bags at room temperature
Bactris gasipaes (peewah) Plate 2	• Can be boiled with salt and other spices and eaten as an appetizer • Can also be preserved in brine	• Harvested using coco-knife, sharp chisel, or a sharp luchette attached to picking rods • Can be stored at room temperatures for up to 1 month • Does not tolerate refrigeration • Spoilage in plastic bags at room temperature
Manilkara bidentata (balata) Plate 3	• Edible fresh fruit • Latex is used industrially for products such as chicle	• Usually harvested in the green to orange stage by picking rods or shaking of trees • Fruits usually have latex which should be allowed to drain • Fruit can be stored up to 1 week at 15°C • Spoilage may occur when left in plastic bags at room temperature

Crop	Utilization	Harvesting and postharvest handling
Passiflora quadrangularis (barbadine) Plate 4 	• Fresh fruit can be made into sherbet or used in desserts • Can also be cooked and used as a vegetable	• The fruits are ready for harvesting when the skin becomes translucent and glossy and is beginning to turn yellowish at the apex • It is clipped from the vine • Very careful handling and packing are essential • Highly perishable
Artocarpus altilis (breadfruit) Plate 5 	• Fruit is usually cooked before eating • Ripe fruit can be eaten as a dessert • Fruit can be boiled, dried, and made into flour, and slices can be fried or stored in brine • The cooked fruit can also be stored • The collected latex can be used as a caulk, glue, and chewing gum	• Fruit stalks must be cut to a length of 1.5 cm and the latex must be drained in the field. • Full green fruits are usually plucked by hand or by a rod. • Fruits can be wrapped in polyethylene and stored at 12°C last about 20 days • Lower temperature causes chilling injury ad higher temperature will allow it to ripen more quickly. • The fruit ripens in 3–7 days • The ideal relative humidity is 90–95% to prevent water and weight loss and shriveling • Well-ventilated cartons should be used and fruits packed in a single layer and separated from adjacent fruits by separators to prevent bruising during transport • Fruit for the local market should also be graded and packed in sturdy crates for delivery • The storage time between packing and delivery to the market should be as short as possible to ensure that the fruit still has a reasonable shelf life when it reaches the consumer
Averrhoa carambola (carambola) Plate 6 	• Fruits can be eaten when fully ripe • Sliced fruits can be added to salads • Can be made into jams, preserves, pickles, candy, juice, and liquor	• Fruit should be firm and crisp with shiny golden-yellow, orange, or yellow skin when ripe. • Fruits are picked by hand • Fruits can be wrapped in tissue paper to avoid rubbing injury • Plastic wraps can also be used • Can be cooled after harvest to extend shelf life • Can be stored up to 5 weeks at 5°C • Moisture loss can occur during storage and can lead to skin browning • Waxing has been reported to be helpful in reducing desiccation

Crop	Utilization	Harvesting and postharvest handling
Syzygium malaccense (pommerac) Plate 7 *Syzygium aqueum* (wax apple) Plate 8 	• The ripe fruit can be eaten fresh or processed into juices	• The skin is thin and delicate and can be easily damaged • Extra care is required when harvesting and handling • Fruits are harvested when the skin has nearly full color and the fruit is firm either by hand or rod • Harvested fruit should be sorted, sized, and packed in a single layer in trays with padding to limit injury • Paper can be used to wrap fruits • The fruit is non-climacteric and is chill sensitive • Fruit can last 4–6 days at ambient temperature • Water loss and sugar content usually declines if fruits are left in polyethylene-wrapped pack at 12°C
Spondias cytherea (pommecythere) Plate 9 	• Harvested either mature-green, semi-ripe, or ripe • It can be eaten fresh or processed into juices or preservatives	• Fruits are harvested when full and light green to yellow by hand or rod • Fruit is climacteric • Fruit must be harvested using the appropriate harvesting tools • Fruits must not be detached from the trees or fall to the ground • Rough handling of harvested fruit must be avoided in order to prevent bruising, cracking, and wounding • Fruit must not be exposed to direct sunlight. They should be kept under shade • Harvested fruits must be transferred to clean and dry plastic crates • Defective fruits, i.e., those that are diseased, mechanically damaged, and otherwise not marketable, must be separated out • Fruit must be cleaned by washing in water containing 100–120 ppm sodium hypochlorite and 0.05% thiabendazole • Surface moisture must be removed by spreading fruit on racks under shade with adequate ventilation • Fruits can be packed into fiberboard waxed cartons • Ambient conditions. For temporary storage under ambient conditions, store in a cool, dry place with adequate ventilation, away from sources of ethylene • Cool storage. Transport and store at 13–15°C and 85–90% relative humidity

Crop	Utilization	Harvesting and postharvest handling
Manilkara zapota (sapodilla) Plate 10 	• When ripe, the fleshy pulp may be eaten or used to make custard and ice cream called "chicle," and it was this—with the addition of massive amounts of sugar—chewing gum	• Fruits should be harvested when the brown scaly external material from the fruit sheds off • Fruit becomes corky brown in color • Ensure that latex does not flow when the fruit is scratched with the finger nail • Maturity usually judged by size and appearance • Harvest with the use of appropriate harvesting tools and ensure that fruit does not touch the ground • Should be carefully handled as fruit can be readily damaged by abrasion and impact • Mature frit ripens in 3–7 days at 25°C and stored at 15°C for 14 days • Lower storage temperatures lead to chilling injury and a failure to soften • Fruit marketability lessened by low "shelf life" and difficulty of grading for similar maturity
Mammea americana L (mamey apple) Plate 11 	• Can be eaten raw in fruit salads jellies or ice cream	• Harvest by hand and avoid the fruit touching the ground • Perishable during transport • Low temperatures required during handling and sensitive to chilling injury • Controlled and/or modified atmosphere packaging extends its storage life
Annona squamosa (sweet sop) Plate 12 	• Edible fresh fruit • Fruit can be pureed and used to prepare iced drink or mixed with other juices, or it can be made into sherbets and gelatin dishes	• Harvest the fruit by hand when it is fully developed and still firm • The color of the fruit peel changes from dark green to pale green • The fruit is still firm when pressed with the thumb or finger • Can be held after harvest for up to 4–7 days at room temperature before softening occurs • The skin of ripening fruit usually turns dark brown to black, but flesh is unspoiled • Harvested fruits must be transferred to cushioned boxes or crates in order to avoid mechanical damage or bruising • Harvested fruit should not be exposed to direct sunlight, but should be kept under shade • Individual fruit should be wrapped in soft packing material and transported in well-ventilated cardboard boxes • Can be kept for up to 7–10 days at 17°C • Pre-cooling is essential to help extend shelf life

Crop	Utilization	Harvesting and postharvest handling
Musa spp. (silk fig) Plate 13 	• Edible fresh fruit or processed to make ice cream	• Light green to yellow fruits should be handpicked • Highly perishable • Can be stored for up to 1–5 days before spoilage occurs • Cannot withstand low temperatures and should not be refrigerated
Psidium guajava (L.) (guava) Plate 14 	• The fruit can be eaten when ripe • Puree can be used to make juice, cakes, sauces, jams, and jellies and ice cream	• Fruit should be handpicked and not allowed to touch the ground • The fruit is usually grown for processing • Highly perishable • Susceptibility to physical damage, chilling injury, diseases, and insect pests • Mature green and partially ripe fruit can be held for 2–3 weeks at 8–10°C • Ripe, soft fruit can be held about 1 week at 5–8°C • RH of 90–95% is recommended [23] • Shelf life is about 7 days when stored at 20°C

Photo credits: Terry Sampson

Table 1. Some of the most common indigenous and underutilized fruits in Trinidad and Tobago, their utilization, harvest, and postharvest practices.

All these tropical fruits offer many diverse aromas, textures, tastes, and shapes and include many different bioactive compounds. These exotic indigenous tropical fruits and root vegetable products are seasonal crops to a great extent and form part of the diaspora niche markets in Europe and North America. However, because of the poor storage qualities and high rate of spoilage due to the poor postharvest handling, the export potential is not realized. There is a gap in the information for the minor exotic crops and fruits for export which have to be transported over a longer period than the local market.

The commercial success of these tropical fruits depends on the development of postharvest technologies and handling techniques both for marketing locally and export. There are many advances in techniques for harvesting, packing, selection and grading, quality evaluation, transportation, and refrigerated storage conditions and management of postharvest physiological symptoms, insect pest, and microbial decay which must be considered.

It has been reported that the major causes of postharvest losses in tropical fruits were due to fungal pathogens particularly *Colletotrichum* spp. [2]. However, the etiology, biology, and environmental and horticultural factors that contribute to the high infection and development of decay are critical in improving fruit quality and must be clearly defined. It has been observed that although tropical fruit crops are cultivated under higher ambient temperatures, they usually have poor postharvest practices, which further affects their shelf life [3]. It was recommended that storage in different packaging materials, viz., micro-perforated polypropylene bags (MPB), micro-perforated polypropylene bags with ethylene absorbent (MPB+K), macro-perforated polypropylene bags coated with antimist coating (PP), fruit waxing, gibberellic acid (GA3), and indole butyric acid (IBA) applications [3].

Many scientific research have been focused on the crop production but little on postharvest issues. However, postharvest quality and shelf life of the fruit in part will depend on some postharvest handling practices and treatments carried out after harvest [4]. Handling practices like harvesting, pre-cooling, cleaning and disinfecting, sorting and grading, packaging, storing, and transportation all play an important role in maintaining quality and extending shelf life. The use of improved postharvest techniques such as refrigeration, heat treatment, modified atmosphere packaging (MAP), and 1-methylcyclopropene (1-MCP) and calcium chloride ($CaCl_2$) has been suggested [4].

This chapter will review and evaluate the applicability and appropriateness of some of these postharvest handling practices and treatment techniques in extending the shelf life and quality and reduction of the losses in some selected indigenous fruits of Trinidad and Tobago.

2. Harvest and postharvest practices and operations for indigenous fruits

Almost all minor tropical fruits are harvested manually, but this physical handling can have a drastic effect on the postharvest quality and shelf life. Most tree ripened fruits are susceptible to rough handling during and after harvesting due to mechanical injuries which reduces postharvest quality and shelf life [5]. The more common handling practices by small farmers include picking, pre-cooling and washing, sorting and grading, wrapping or packaging, transport, and storage.

Small-scale producers throughout the Caribbean usually harvest most of their fruits very early, when fruits are sometimes green to avoid the incidence of praedial larceny (farm theft) [6]. This often results in underripe fruits which have not yet developed their full flavor. Mechanical damage during harvest is therefore not often a serious problem.

The maturation of fruit ready for harvest varies depending on the fruit type which may be harvested in either the matured green, partially ripe, or ripe state [7]. The physiological stages of fruit maturity are the maturation, ripening, and senescence, and they influence the postharvest quality and shelf life. The climacteric fruits are usually harvested at the matured green and then placed in special conditions that ensure their continuation into ripening and senescence. There is little fruit loss during this phase except with physical damage or fruit-fly infested fruits.

The fully ripened fruits are more susceptible to mechanical injuries during harvesting and should be harvested during a cool period of the day to avoid excessive field heat accumulation. The use of sharp tools or containers with edges is discouraged as these bruise and puncture the fruits (**Table 1**).

3. Recommended best practice postharvest treatments

3.1. Pre-cooling and field heat reduction after harvest

There is usually a high build of field heat in the afternoon which is undesirable during the harvesting stage. This increases the metabolic activity that should be reduced immediately after harvest. Pre-cooling minimizes the effect of microbial activity, metabolic activity, respiration rate, and ethylene production [8, 9].

Pre-cooling treatments can reduce the ripening rate, water loss, and decay, thereby preserving quality and extending shelf life of most fruits. It is critical to determine the specific temperature range usually between 13 and 20°C for specific fruits. Pre-cooling can be done by hydro-cooling or dipping fruits in cold water with disinfectants such as thiabendazole and sodium hypochlorite to reduce microbial infections [10].

3.2. Sanitization

Proper hygiene can prevent postharvest diseases and transmission of food-borne pathogens such as *Salmonella, Cryptosporidium, Cyclospora,* and hepatitis A virus particularly where water is a constraint or likely to be contaminated. Care should be taken to ensure that harvested fruits do not come into contact with the ground to avoid any source of contamination. Disinfection or sanitization in conjunction with hydro-cooling is recommended using sodium hypochlorite solution to reduce the incidence of fungal infection before any postharvest process [11]. Thiabendazole [12] and anolyte water dips [13] have been shown to reduce the microbial infection and maintain superior fruit quality during storage.

3.3. Grading and sorting

Sorting and removal of damaged, infected, or diseased fruits can prevent the production of ethylene in substantial amounts which is capable of affecting the adjacent fruits. When sorting is followed immediately by grading and categorizing of fruits based on color, size, stage of maturity, or degree of ripening, cross contamination and the spread of infectious microbes from discarded fruits to healthy fruits during postharvest handling are reduced. Additionally, through removal, overripen fruits reduce ethylene production and reduce "forced" ripening of the other fruits.

3.4. Packaging

Packaging is done to improve handling, storage, and transport. It involves enclosing the product to protect it from mechanical injuries, tampering, and contamination from physical, chemical, and biological sources [14]. The materials used vary from wooden and plastic crates, cardboard and styrofoam boxes, woven palm baskets to nylon, jute sacks, and polythene bags. However, not all materials give full protection to the fruits [15].

Improper or inadequate aeration within the packaged commodity can result in a buildup of heat as a result of respiration. Similarly, crates or baskets with rough surfaces and edges can induce mechanical injuries which will increase ethylene production, thus reducing postharvest quality of the fruits [5].

3.5. Storage

Tropical fruits usually have a very high moisture content and are very difficult to store at ambient temperatures for extended periods without refrigeration. It is critical that storage is extended so that processing operations can provide continuity of product supply throughout the seasons.

Most fruits can be stored for short terms at ambient temperature (10–15°C) and relative humidity (85–95%) [16], provided that ventilation is adequate to prevent heat built from respiration and reduce poor ripening and chilling injuries. However, these conditions in tropical countries are difficult to achieve due to high day temperatures and high relative humidity [17]. On the contrary, very low storage temperature is also detrimental to the shelf life and quality of many tropical fruits and will reduce its flavor, total soluble solids (TSS), and pH of the fruit [18].

It is essential that the correct temperature management is predetermined during storage so as to extend the shelf life of the fruit while maintaining fruit qualities. Storage for short to intermediate time by using evaporative cooling system is recommended.

3.6. Transportation

Farms are usually located away from the marketing centers, processing plants, and inaccessible roads. The lack of proper transportation, like refrigerated vehicles, is a big challenge for both producers and distributors in Small Island Developing States (SIDS). This causes unnecessary delays in getting the produce to the pack houses or market, and such delays can result in losses up to about 20%. During transportation, the produce should be immobilized by proper packaging and stacking to avoid excessive movement or vibration or heat buildup, if transport is not refrigerated. Refrigerated trucks are not only convenient but also effective in preserving the quality of fruits. But it has a high initial investment which is usually unaffordable in SIDS.

4. Selected postharvest treatment techniques

After harvesting, fruits remain living, and all the metabolic process and functions of living tissue continue. However, the postharvest quality of the fruits at or after harvest cannot be enhanced by any postharvest technology, except fruit color, but can be maintained or prolonged.

4.1. Refrigeration storage

Refrigeration is one of the most common effective methods of preserving the quality of many fruits for several days. Low fruits placed at storage temperature can protect quality attributes like texture, nutrition, aroma, and flavor in many harvested fruits and extend shelf life. However, some fruits are sensitive to chilling injury when they are stored below their critical temperature (10°C) for even short durations and may result in pitting, uneven ripening, and fungal infestation of stored fruits.

The required optimum temperatures (10–15°C at 85–95% relative humidity) can be achieved by using less expensive methods of cooling such as an evaporative cooling system [19]. Where, air temperatures can be decreased (16°C, 91% relative humidity) without significant deterioration due physiological weight loss.

4.2. Postharvest heat treatment

Postharvest heat treatments using hot air and heated water have been reported to reduce chilling injuries in fruits like mangoes and oranges. Heat treating (37–42°C) prior to cold storage can slow down ripening while increasing pathogenic resistance enhanced or caused no change in some quality traits particularly in modified atmosphere storage system. Heat treatment before cold storage at 14°C can also increase TSS and titratable acids (TA) when fruits ripened as compared to the untreated fruits.

4.3. Modified atmosphere packaging (MAP)

Modified atmosphere packaging (MAP) is a packaging technique of using packaging materials with predetermined composition of gases (O_2 and CO_2) after which there is no active effort of modifying the storage space. The materials allow for diffusion of gases through them until a stable equilibrium is reached between the external gases and those inside the package.

MAP materials used include polyethylene terephthalate (PET), low-density polyethylene (LDP), high-density polyethylene (HDP), polyvinyl chloride (PVC), polypropylene [20], and polystyrene [21, 22].

MAP provides a modified atmosphere to control ripening, reducing water loss in stored products reducing mechanical injuries, and reduces the spread of food-borne diseases [23]. The water loss and subsequent shriveling of fruits in tropical regions are two of the causes of their deterioration. However, maintaining excessively high levels of relative humidity inside the package can result in moisture condensation on the commodity, which will in turn create a conducive environment favorable for pathogenic activities, thus increasing the risk of fruit deterioration.

4.4. Methylcyclopropene (1-MCP)

Treatments of 1-methylcyclopropene (1-MCP) can suppress the action of and reduce the rate of ethylene production in harvested climacteric fruit [24]. Fruits with high metabolic activities usually have a shorter shelf life, and any attempts to slow down the metabolism will increase the shelf life. The reduced metabolic activities are closely associated with the ripening process such as color change, cell wall breakdown, and respiration rates and are advantageous to extending storage life [24] and the prevention of abscission in fruits [25].

4.5. Calcium chloride (CaCl$_2$) application

The pre- and postharvest treatment of calcium chloride (CaCl$_2$) has been shown to improve shelf life and quality of many fruits [26] by delaying ripening and senescence, reducing respiration, maintaining firmness, and reducing physiological disorders [27, 28].

It has also been shown to reduce fungal and other microbial infections and maintain the structural integrity of cell walls, delaying softening and extending shelf life by 4–5 weeks [29]. It also aids in maintaining the quality of fruits by reducing the physiological disorders, increasing the fruit firmness, delaying ripening process, and prolonging the shelf life [30]. Fruit color loss is also delayed, and ethylene production is reduced by 92% [26] while maintaining higher firmness levels during storage. CaCl$_2$ is a very cheap and cost-effective soluble salt which can be dissolved easily and used in the pre-cooling or cleaning of the fruits after harvesting.

Author details

Puran Bridgemohan[1] and Wendy-Ann P. Isaac[2]*

*Address all correspondence to: wendy-ann.isaac@sta.uwi.edu

1 Faculty of Biosciences, Agriculture and Food Technologies, The University of Trinidad and Tobago, Trinidad and Tobago

2 Department of Food Production, Faculty of Food and Agriculture, The University of the West Indies, St. Augustine, Trinidad and Tobago

References

[1] Muhammad RH, Bamisheyi E, Olayemi FF. The effect of stage of ripening on the shelf life of tomatoes (*Lycopersicon esculentum*) stored in the evaporative cooling system (E.C.S). Journal of Dairying, Foods & Home Sciences. 2011;30(4):299-301

[2] Yahia EM, De Jesus Ornelas-Paz J, Gonzalez-Aguilar GA. Nutritional and Health promoting Properties of Tropical and Subtropical Fruits. In: Postharvest Biology and Technology of Tropical and Subtropical Fruits. Edited by Yahia EM: Woodhead Publishing; 2011. pp. 21-78. ISBN 9781845697334. https://doi.org/10.1533/9780857093622.21

[3] Droby S, Wisniewski ME, Benkeblia N. Postharvest pathology of tropical and subtropical fruit and strategies for decay control. In: Yahia E, editor. Post-harvest Biology and Technology of Tropical and Sub-tropical Fruits. Vol. 1. Cambridge, UK: Woodhead Publishing Limited; 2011. pp. 194-223

[4] Arah IK, Ahorbo GK, Anku EK, Kumh EK, Amaglo H. Postharvest Handling Practices and Treatment Methods for Tomato Handlers in Developing Countries: A Mini Review, Advances in Agriculture, vol. 2016, Article ID 6436945, 8 pages, 2016. DOI:10.1155/2016/6436945

[5] Arah IK, Amaglo H, Kumah EK, Ofori H, Preharvest and Postharvest Factors Affecting the Quality and Shelf Life of Harvested Tomatoes: A Mini Review. International Journal of Agronomy, vol. 2015, Article ID 478041, 6 pages, 2015. DOI:10.1155/2015/478041

[6] Isaac WAI, Ganpat W, Joseph M. Farm security for food security: Dealing with farm theft in the Caribbean Region. In: Ganpat W, Dyer R, Isaac Wendy-Ann P., editors. Agricultural Development and Food Security in Developing Nations. Pennsylvania USA: IGI Publications; 2017. (ISBN 978-152-250-942-4). pp. 300-319

[7] Moneruzzaman KM, Hossain ABMS, Sani W, Saifuddin M, Alenazi M. Effect of harvesting and storage conditions on the postharvest quality of tomato (*Lycopersicon esculentum* Mill) cv. Roma VF. Australian Journal of Crop Science. 2009;3(2):112-121

[8] Akbudak B, Akbudak N, Seniz V, Eris A. Effect of pre-harvest harpin and modified atmosphere packaging on quality of cherry tomato cultivars "Alona" and "Cluster". British Food Journal. 2012;114(2):180-196. https://doi.org/10.1108/00070701211202377

[9] Shahi NC, Lohani UC, Chand K, Singh A. Effect of pre-cooling treatments on shelf life of tomato in ambient condition. International Journal of Food, Agriculture and Veterinary Science. 2012;2(3):50-56

[10] Ferreira MD, Brecht JK, Sargent SA, Aracena JJ. Physiological responses of strawberry to film wrapping and precooling methods. Proceedings of Florida State Horticulture Society. 1994;107:265-269

[11] Genanew T. Effect of postharvest treatments on storage behaviour and quality of tomato fruits. World Journal of Agricultural Sciences. 2013;9(1):29-37

[12] Batu A, Thompson AK. Effects of modified atmosphere packaging on post harvest qualities of pink tomatoes. Journal of Agriculture and Forestry. 1998;22:365-372

[13] Arjenaki OO, Moghaddam PA, Motlagh AM. Online tomato sorting based on shape, maturity, size, and surface defects using machine vision. Turkish Journal of Agriculture and Forestry. 2013;37:62-68

[14] Prasad P, Kochhar A. Active packaging in food industry: A review. IOSR Journal of Environmental Science, Toxicology and Food Technology. 2014;8(5):01-07

[15] Idah PA, Ajisegiri ESA, Yisa MG. An assessment of impact damage to fresh tomato fruits. Australian Journal of Technology. 2007;10(4):271-275

[16] Žnidarčič D, Trdan S, Zlatič E. Impact of various growing methods on tomato (*Lycopersicon esculentum* Mill.) yield and sensory quality. Research Reports Biotechnical Faculty, University of Ljubljana, Agriculture. 2003;81(2):341-348

[17] Parker R, Maalekuu BK. The effect of harvesting stage on fruit quality and shelf-life of four cultivars (*Lycopersicon esculentum* Mill). Agriculture and Biological Journal of North America. 2013;4(3):252-259

[18] Moretti CL, Sargent SA, Huber DJ, Calbo AG, Puschmann R. Chemical composition and physical properties of pericarp, locule and placental tissue of tomatoes with internal bruising. Journal of the American Society for Horticultural Science. 1998;123(4): 656-660

[19] Workneh TS, Woldetsadik K. Forced ventilation evaporative cooling: A case study on banana, papaya, orange, mandarin, and lemon. Tropical Agriculture. 2004;81(1):1-6

[20] de Wild HPJ, Otma EC, Peppelenbos HW. Carbon dioxide action on ethylene biosynthesis of preclimacteric and climacteric pear fruit. Journal of Experimental Botany. 2003;54:1537-1544

[21] Artés F, Gómez P, Artés-Hernández F. Modified atmosphere packaging of fruits and vegetables. Stewart Postharvest Review. 2006;5:1-13

[22] Sandhya. Modified atmosphere packaging of fresh produce: Current status and future needs. LTW-Food Science and Technology. 2010;43:381-392

[23] Kader AA, Watkins CB. Modified atmosphere packaging – Towards 2000 and beyond. Hort Technology. 2000;10(3):483-486

[24] Watkins CB. Overview of 1-methylcyclopropene trials and uses for edible horticultural crops. HortScience. 2008;43:86-94

[25] Passam HC, Karapanos IC, Bebeli PJ, Savvas D. A review of recent research on tomato nutrition, breeding and post-harvest technology with reference to fruit quality. The European Journal of Plant Science and Biotechnology. 2007;1:1-21

[26] Senevirathna PAWANK, Daundasekera WAM. Effect of post-harvest calcium chloride vacuum infiltration on the shelf life and quality of tomato (cv. 'Thilina'). Ceylon Journal of Science (Biological Sciences). 2010;39(1):35-44

[27] Hong MN, Lee BC, Mendonca S, Grossmann MVE, Verhe R. Effect of infiltrated calcium on ripening of tomato fruits. LWT Journal of Food Science. 1999;33:2-8

[28] Akhtar A, Abbasi AA, Hussain A. Effect of calcium chloride treatments on quality characteristics of loquat fruit during storage. Pakistan Journal of Botany. 2010;42(1):181-188

[29] Lara I, García P, Vendrell M. Modifications in cell wall composition after cold storage of calcium-treated strawberry (Fragaria×ananassa Duch.) fruit. Postharvest Biology and Technology. 2004;34:331-339

[30] Abbasi AA, Zafar L, Khan HA, Qureshi AA. Effects of naphthalene acetic acid and calcium chloride application on nutrient uptake, growth, yield and postharvest performance of tomato fruit. Pakistani Journal of Botany. 2013;45(5):1581-1587

9

Post-Harvest Handling of Freshwater Crayfish: Techniques, Challenges and Opportunities

Japo Jussila, Ravi Fotedar and Lennart Edsman

Additional information is available at the end of the chapter

Abstract

Development of post-harvest handling practices of crayfish in Fennoscandia has largely been based on their high value and has historical rationale. Crayfish were transported from Finland to St. Petersburg and Central European markets already in the early mid-1800s using rather sophisticated methods. Crayfish require cool and moist environment for stress minimisation and may easily be transported and stored to live out of water if these principles are followed. During post-harvest process, it is important to minimise handling of crayfish from the point traps which are pulled until crayfish are processed for food. Crayfish should be protected against the elements and stored in cool containers during on board and land transport, after initial sorting of the catch. In the holding depot, crayfish are normally sorted for the second time and then placed in holding facilities waiting for the transport to markets. Holding facilities could be tanks or more developed and cost-effective systems, such as CrayShower. This storage system is based on crayfish being stored out of water in moist environment. The main principles during post-harvest handling are to keep physical disturbances to the minimum and to provide cool and moist environment for crayfish.

Keywords: catch, stress, survival, mortality

1. Introduction

1.1. Crayfisheries and catch in the Fennoscandian countries: history and present time

Freshwater crayfish have since ages formed a crucial part in the Fennoscandian culture both economically and culturally [1–4]. Finnish tenant farmers used crayfish to fulfil their obligations with their landlords, as part of their annual work quota which could be substituted by providing crayfish instead of labour [5]. This greatly helped since the kids could easily contribute to the well-being of the family by trapping the required crayfish. Since then, the productive wild stock of crayfish has allowed the purchase of the first work horses, bikes, mopeds, cars and even elevation of the standard of living in the Fennoscandian countryside [5–7].

During the turn of the 1900s, crayfish exports from Finland to St. Petersburg region peaked around 16 million individual crayfish [1, 6], and this trade has been claimed to have been a corner stone for even initiation of industrial-scale enterprises in Finland. There was also a trade of crayfish to both directions between Finland and Sweden, and this trade is claimed to have introduced crayfish plague (i.e., *Aphanomyces astaci*, the disease agent) into Sweden [4, 8]. Because of the nature of the Finnish and Swedish aquatic ecosystems, there have traditionally been only a limited number of Finns and Swedes not being aware of the noble crayfish and tradition of crayfish trapping. Well, the situation has changed dramatically recently.

Crayfish plague epidemics ended the golden age of the noble crayfish fishery and trade in Fennoscandia around 1910, after the first epidemics were observed in 1893 in Lake Saimaa in Finland and some 14 years later around Stockholm in Sweden [4–6]. Regardless, the tradition of trapping crayfish and selling them to be enjoyed during autumn crayfish parties is still strong. Catches are lower and prices are higher [1, 2], but there is still a strong demand for delicious freshwater crayfish in Finland and Sweden. Thus, the development of better means to ensure maximum catch is required, and post-harvest handling methods have improved.

Crayfish trapping and post-harvest handling of the catch are closely connected to conservation of valuable native crayfish stocks [3]. For crayfish trapper, it is of utmost importance to remember that crayfish, once taken out of the water and transported away from the trapping site, should never be put back to natural water bodies again. This is to ensure that crayfish would not be accidentally introduced to new water bodies and, more importantly, the diseases accompanying crayfish would not spread. The crayfish plague has been proven a big kill joy!

1.2. Value of crayfish

Freshwater crayfish have traditionally had a high commercial value in Europe [5, 6], which has encouraged trapping and trade of the valuable natural resource. Crayfish have even been used as an alternative for currency, which was greatly improving the standard of living especially in the countryside and among rural population in Finland and wider in Fennoscandia, too [6, 8]. The need to develop post-harvest handling practices surfaced early with holding of crayfish along water bodies in cages and transport over short or long distances in special crates. Especially in Finland, reaching the Russian and European markets [6] required gentle post-harvest handling of crayfish. These inventive systems could be afforded and applied due to value of crayfish catch.

As an example, during the turn of the twentieth century, crayfisherman could be payed from 1 to 2 Finnish penni for a crayfish, while an industrial workers' pay was from 3 to 4 Finnish markka per day [6]. Thus, a crayfisherman having access to a productive noble crayfish stock could easily earn equal to industrial worker's pay during crayfish trapping season. Later, for example, during the 1980s and 1990s, the monetary value of the freshwater crayfish in Finland was roughly a minimum of 10 FIM each, and the beach price has been during the 2000s roughly from 1.50 to 2 € per 10 cm TL noble crayfish [1]. As the trade is based on very basic capitalistic principle, the price has varied somewhat depending on supply and demand in addition to various sociological issues. In Sweden, the price is based on weight, not on individual crayfish. The beach price has ranged between 290 and 530 Swedish krona per kilo (i.e., about 30 and 55 €) for noble crayfish and between 105 and 220 Swedish krona (i.e., about 11 and 23 €) for signal crayfish. Thus, crayfish have traditionally been considered as delicacy, especially when catch has been low. The scenario has somewhat changed due to the introduction of signal crayfish into the wild in Fennoscandia, and the price for the signal crayfish has been lower than for the noble crayfish. Even with the lowered price for the signal crayfish, crayfish are still considered gourmet food for festivities [9].

1.3. Past aspects of freshwater crayfish post-harvest handling

It was a common knowledge right from the start of crayfish trade in Finland that crayfish survive out of water under certain conditions [6]. This allowed a lively trade of crayfish, as they could be transported for long distances. Crayfish were thus highly valued, as the rural population could exploit this natural resource, supply even Central European market, and greatly improve their living standards.

Traditionally, crayfish have been transported both on board and on land for long distances in Fennoscandia, with some of the crayfish been trapped in Häme County (Finland) and Stockholm County (Sweden) transported for markets of St. Petersburg, Hamburg and even Paris as early as 1910 [10, 11]. Even a narrow railroad track was built from Lake Erken to the coast for the sole purpose of transporting crayfish catch to be shipped further by boat to other cities in Sweden and the continent [11, 12]. It was discovered that crayfish can survive lengthy periods out of water, if kept under proper conditions and processed as soon as they reached their destination. Crayfish were also placed in submerged cages during the transport, if overnighting was required. This eased their transportation stress but also allowed a rapid spreading of the diseases [6] and also crayfish, as they sometimes escaped from the cages.

Crayfish have traditionally been first stored in wooden cages around the shores of the water bodies where they have been trapped [6]. The wholesalers have then been collecting the catch from crayfish trappers and transported them either to the local markets or further to export markets in baskets made from wood splints lined with moist moss. During transport, conditions might have worsened, especially temperature could have risen, and caused stress resulting in increased mortality. Until recently, crayfish have been stored under similar conditions during transport, i.e., open cardboard or plastic boxes or crates have been used with crayfish been covered with moist cloth or layer of the tree, for example, alder and branches [10].

Crayfish have been collected by the wholesalers to be marketed and held first in cages along the shoreline of the lakes and later in holding depot's tanks [6]. Due to aggression [13, 14], this communal holding could have resulted in losses, as crayfish might have been moulting in addition to aggressive behaviour caused by crowding.

The increased mortality has traditionally been largely accepted as part of the risks included in the transport of crayfish catch with little or no development of more suitable applications for transport. One of the reasons for the lack of development has been the fact that large proportion of crayfish catch in Fennoscandia has been trapped by recreational trappers working with small-scale catches and respecting traditional aspects of crayfish trapping [1].

2. Factors affecting post-harvest handling procedure

Stress is an individual's response to challenge, and if the condition is prolonged, it can affect the well-being of the individual crustacean [15, 16]. Any diversion from the optimum conditions can cause stress [17], and the avoidance of stress is essential for the best practices during post-harvest handling of crayfish, as stress causes losses among the catch both as elevated mortality and as quality of the individual crayfish. In addition to the normal maintenance of the well-being of animals, crayfisherman and wholesaler have to focus on maximum cost-effectiveness during trapping and post-harvest handling processes. The elevated individual mortality among the catch as such is a loss for crayfisherman, but it also is prone to create circumstances under which the overall well-being of the whole catch might be declining due to worsening of transport and holding conditions. Moribund and dead crayfish cause water quality problems especially when crayfish are held communally and they share the same water body. Furthermore, the price of crayfish, and thus crayfishermen's income, relies on the quality of the catch, which is largely based on the capability of crayfish to survive through the marketing chain to the consumers in prime condition.

Freshwater crayfish can survive out of water for long periods, even days [10, 18], which allows several options for both transport of the catch and holding for commercial purposes. The optimisation of the post-harvest handling conditions is essential in order to minimise handling stress and mortalities [19–22]. Crayfish should be handled as briefly and gently as possible, kept moist and cool all the time [16, 18].

Crayfish should remain in wet and cool transportation and handling environment throughout the post-harvest handling chain [10, 16]. The gill structure allows gas exchange as long as the surface layer of the gills remains moist, because, contrary to fish, the gill filaments do not collapse when not supported by the ambient water, thus allowing close to normal function even out of water. Cool conditions on the other hand minimise evaporation and lower metabolic rate, thus enabling resource allocation to prevent detrimental effects of the post-harvest handling stress.

The moisture level should be as close as possible to 100% humidity during every part of the post-harvest handling chain (e.g., Refs. [10, 23]). This enables gas exchange through gills and lowers stress, as crayfish would not be experiencing water loss through surface tissues. During the post-harvest handling, the ambient temperature should remain below 20°C, preferably even closer to 10°C, to slow down the metabolism of crayfish and to pacify them, as the transport mortality is prone to decrease with the lowering temperature. In addition, the temperature fluctuation should be kept minimal, since the temperature changes, especially rapid and significant changes within the upper optimum limit, could be causing stress and require resources from crayfish [24]. The catch has to be protected against direct sunlight, as there are multiple effects of temperature elevation, evaporation and intensive light causing stress and worsening post-harvest handling conditions.

Air exposure during transport could cause severe dehydration, even though weight loss between 3.9 and 4.5% during transportation without any specific negative effects has been reported [23, 25]. Some of the water is lost from gill chambers, but also haemolymph has been suggested to dehydrate. The slight dehydration does not seem to cause elevated mortality, especially if the transported crayfish are later submerged.

Moulting causes a drop in the catch and affects the well-being of crayfish via physiological changes during post-harvest handling processes (e.g., Refs. [14, 26]), since crayfish would be more susceptible to stress close to a moult. Normally, there are at least two periods of moulting during trapping season in Fennoscandia, when commercial catch declines and the survival of crayfish declines, too. Both of these factors affect the economics of the trapping, and special care should be taken during the moult-related changes in the condition of crayfish. While handling, crayfish can be detected as newly moulted or approaching moult by the texture and hardness of the carapace. Dead or moribund crayfish could worsen the conditions for the surviving crayfish if held communally, due to possible damage of tissues and release of microorganisms. Post-harvest handling and transport of those crayfish close to the moult should thus be avoided.

Crayfish require shelter during both transport and holding. The availability of shelter pacifies crayfish and thus reduces stress [27]. Shelter also provides refugee during communal holding, when crayfish are held in tanks in the holding depots. Shelters can be designed to allow easy collection of crayfish from the tanks, since they tend to spend most of their time in shelters to avoid aggressive interactions with co-species and exposure to ambient conditions, such as light. Shelter during transport, on the other hand, is prone to reduce stress, as crayfish will be offered escape from light, dry conditions and elevated temperature.

Predatory pressure during communal holding can result in both excess mortality and aggressive behaviour [13]. During the growth season and if there are odd moulting events in the holding tanks, the mortality can be elevated and decaying crayfish affect the water quality in the tanks negatively. This can be avoided by storing crayfish in low densities, sorting of the catch by size and providing adequate shelter in the tanks. All of these methods may be costly and thus recommended only if the tank holding of crayfish is the only option available. On the other hand, sorting of the catch for commercial purposes, for example, by size, could later decrease the need for handling prior to transit to markets and thus would be a benefit.

Harvesting practices should be planned in order to minimise stress of the market quality crayfish [16], as well as those to be returned back to water as belonging to side catch. The commercial proportion of the catch has to be transported to markets in premium condition, and this part of the catch should experience minimal stress. Crayfish from the side catch should not be excessively stressed by the process being caused by pulling of traps and sorting of catch, thus increasing their chances to survive and grow to commercial size. Practices on board and during transport dictate the fate of the commercial catch, as stress is prone to cumulate and initiation of the stress can be a point of no return resulting in increased mortality during post-harvest handling. Thus, commercial part of the catch is the main priority and should be handled first, while crayfishermen should remember that side catch crayfish are future commercial catch.

Exposure to adverse elements, causing stress, is the main cause of losses during post-harvest handling of crayfish. There are several individual factors stressing crayfish, in addition to the main ones discussed here. Furthermore, the combination of different stressors can be detrimental. Thus, focusing on minimisation of stress of any kind and combination is crucial.

3. Boat transfer: from the traps to the shore

3.1. General handling of crayfish after traps have been pulled

The general rule is that the handling of crayfish should be kept to the minimum after the traps have been pulled. Crayfish should also experience minimum exposure to the elements while on board, as both of these cause stress to crayfish and can result in losses during boat transport and afterwards. The moment when the traps are pulled is crucial, because handling of crayfish at this point of time reflects to their future stamina and thus also survival (e.g., Ref. [16]). The process of handling and transport is prone to cause stress regardless of crayfish trappers' handling and management practices, but the level of disturbance can be controlled and stress minimised by following best practice of post-harvest handling. This point of stress initiation is thus crucial, and it cannot be overly emphasised that the fate of the catch is set at this point of time.

Crayfish should be sorted and stored according to the marketing criteria right after the traps have been pulled. Normally, the catch is sorted according to the size and general appearance. Those crayfish targeted for the further post-harvest handling and finally the commercial market should be stored according to standards aimed for the minimisation of stress immediately after sorting [28].

If the catch has to be stored any lengths of time exposed to sun, wind or dry air, crayfish have to be kept moist and as cool as the conditions on board allow [16]. Conditions on board should allow storing crayfish in the final transport conditions immediately. Crayfish can be sprayed with water, if kept in open containers, as the sprayed water improves holding conditions and could decrease temperature for the catch.

The key principle of crayfish handling after the traps have been pulled is to minimise handling time in order to avoid stress.

3.2. Sorting of crayfish on board

Sorting of the catch on board has become a routine among some of the commercial crayfish trappers in Fennoscandia (**Figure 1**). This allows for minimising handling-related stress for the commercial catch. In the optimum case, crayfish could then be handled next time in the holding depot during a more detailed, additional sorting.

Figure 1. Sorting of crayfish catch on board according to commercial and side catch criteria. White crates with partial lids for the commercial catch and grey crate for temporary holding of the side catch.

Crayfish catch can be sorted by size, by appearance, by sex and sometimes even by species [29]. Quite often, if the catch is sorted on board, only a rough sorting to potential commercial catch and those crayfish to be released is carried out. In Sweden, specific grid systems for automatic sorting of crayfish catch by size have been utilised (**Figure 2**). The sorting is happening after a

Figure 2. Sorting of crayfish catch on board by size. A system of a grid with expanding distances of metal bars allows crayfish to be sorted by size as they slide on the grids. Photo by MSc Fredrik Engdalh (SLU Aqua, Sweden).

set of traps is pulled, depending on crayfish trapping practice of individual crayfishermen. An advanced commercial crayfish trapper could have from 10 to 20 traps in one set of traps, joined by a long line. This set of traps is then pulled and emptied, rebaited and set back, before crayfisherman sorts the catch. With a deckhand, sorting could happen when the traps are pulled, and thus the time the catch is exposed to the elements is minimised; such practice would also minimise the stress. The sorted catch is then stored appropriately to allow stress-free on board transport.

3.3. Holding conditions during boat transport

It is essential to provide cool, moist and shady conditions for the boat-transported crayfish, with special care to avoid exposure to sun [17]. As crayfish will be experiencing some form of unavoidable physical disturbance, i.e., shaking of the boat due to waves and general movement around a lake or river, other forms of stress have to be minimised.

Crayfish tend to go passive with lowering of the temperature. It is preferential to provide stable conditions during transport, and for this purpose, a simple environment within cooled foam container, i.e., esky or foam box for fish, has shown to be optimal [10, 23] and is commonly used both in Sweden and Finland (**Figure 3**). The instant cooling of crayfish under conditions, which also provide immobilisation and moisture, has shown to decrease stress [16, 28] and minimise mortalities even during prolonged transport.

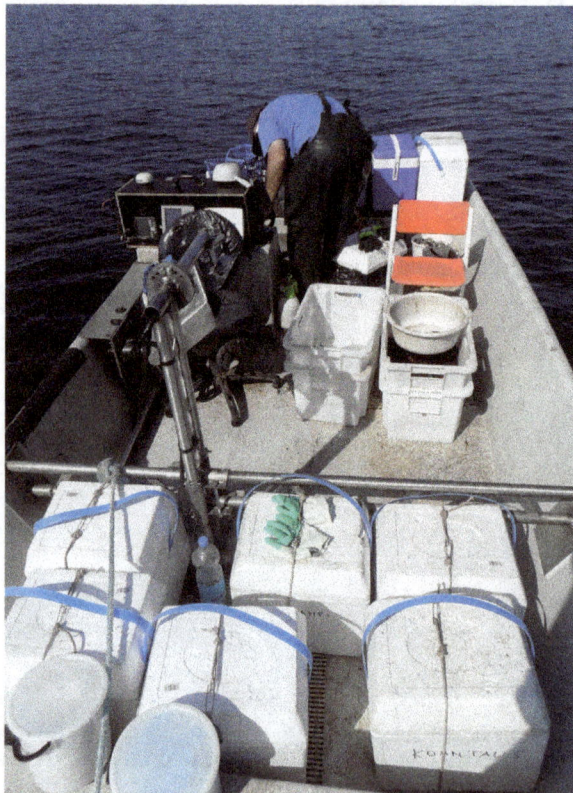

Figure 3. Crayfish transported on board in foam boxes with cooling and moisture provided.

The foam boxes are inexpensive, provide efficient insulation and are easy to handle. Cooling units are placed on the bottom of the foam box, then a plastic or rubber grid is placed on the cooling units to prevent crayfish from frostbites and crayfish are placed inside a plastic bag to ensure proper moisture level and to prevent evaporation and sun exposure during handling on the catch (**Figure 4**). This method has been shown to minimise mortality during transport [10] with cool and passive crayfish being easy to handle later, too, if needed.

Ventilation of the transport boxes and containers is not needed [30], since there is plenty of oxygen in the air and gas exchange does not elevate the metabolic gas levels to those causing stress to crayfish. The ventilation holes could actually worsen the conditions inside the transport containers, as ventilation could increase the temperature inside the box and lower moisture level, too.

Crayfish can be transported on board in plain plastic containers, while this normally does not allow proper cooling and exposes crayfish to the elements. For short on board transport distances and weather conditions not challenging crayfish, this very basic method could be sufficient. A moist fabric on top of crayfish would then allow for shade, some cooling and moisture. These rather hardy plastic containers are sold in local supermarkets, and they can be durable, thus their popularity.

An on board CrayShower application has also been tried by the Lake Saimaa crayfishermen (**Figure 5**). The system is based on constant watering of the catch using fresh surface water from the lake. The CrayShower application requires outside energy source for pumping of the water and is dependent of the surface water temperature. The system works efficiently providing that at least the water temperature remains low, below 20°C, although during hot periods in summer, the sprayed water does not provide sufficient cooling.

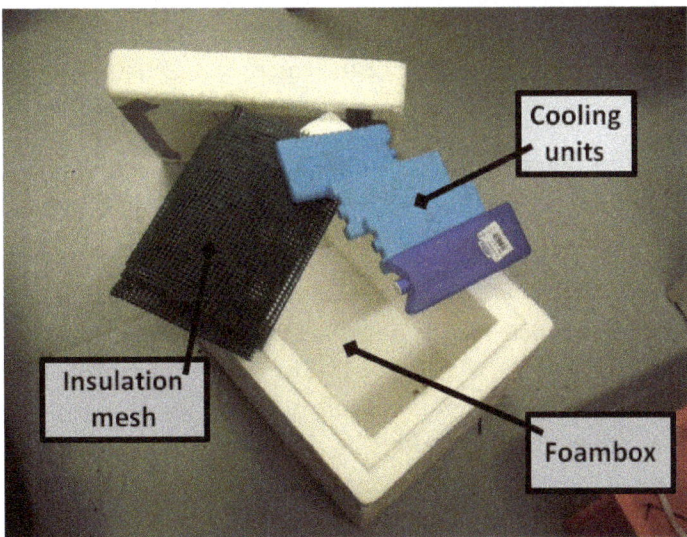

Figure 4. A simple and cost-efficient method for transporting crayfish on board and on land [10] with explanation on different parts inside the cooling box (an additional large plastic bag needed for protection of crayfish).

Figure 5. CrayShower on board transport application. This system uses lake surface water, thus not suitable during warmest summer days.

4. Land transfer: from the shore to the holding depot

The transport on land should also be planned to minimise stressful conditions and rough handling of the catch. Especially during warm summer months, it could easily happen that the conditions get worse, i.e., warmer, during the land transport. Again, it would be beneficial if the number of different handling phases of the transported crayfish could be kept to a minimum and thus disturbance from handling of the catch, too. Crucial for stress minimisation is that crayfish are immobilised and thus do not react easily to gentle handling.

It would be beneficial to use the same containers on board of the boat and during land transport. This would be in line with the demand for minimal handling. It would also be beneficial if specific containers with cooling and immobilisation of the catch have been used on board.

During land transport, the catch has to be protected against heat, evaporation and excess light. Normally, the catch is taken from the boat directly to the transport vehicle, which could be a trailer or a truck. The trailer should have a cover to allow for proper protection. A truck could also be equipped with cooling system [16, 31], which would further improve transport conditions. Regardless of the transport method on land, the containers where the catch is held have to be secured to prevent excess shaking.

It would be preferential to allow the catch to remain in their original containers, untouched, until the holding depot is reached and further processing of the catch is taking place [10]. Even at this point, the catch should be handled gently to avoid activating individual crayfish, which may cause aggressive behaviour and harm to the less aggressive crayfish in the catch.

5. Holding of crayfish: simple is beautiful

Conditions at holding depot have to be planned to minimise handling of crayfish and thus to minimise stress at this point, too. Once again, the less crayfish have to be handled, the better. The attempt is to explore the markets to ensure that the catch can be sorted and stored under conditions, which allow minimal handling when preparing crayfish for transport to markets or next level of wholesalers.

Normally, the catch has to be sorted according to the market demand at the holding depot. The standard criteria for sorting include variables such as size, intact claws and visible trauma [29]. The first sorting on board or during trap pulling rids the most obvious substandard crayfish, but depending on the conditions and handling routine, another sorting may be required. The sorted crayfish will then be stored in separate, graded as per different orders or quality standards. Again, the sorting has to be carried out with minimal stress for crayfish, since they have already experienced stressful conditions during transport.

Crayfish have traditionally been held in tanks in the holding depot with basic freshwater crayfish aquaculture principles applied regarding conditions [14]. The communal holding of crayfish requires good conditions, as the method itself is prone to cause stress [32]. The water quality has to be optimal and water temperature low to minimise mortality during holding, due to the risk of worsening conditions resulting in elevated or mass mortalities.

Normally, the tanks are simple, and no hides are provided to allow proper water flow in all parts of the tank and to enable efficient cleaning of wastes [14, 16]. Quite often tanks are covered to allow shade for crayfish. Water level can be low, less than 20 cm, but the water exchange has to be high to ensure optimum water quality. Due to the aggressive nature of crayfish, cannibalism is commonplace even if crayfish are being fed.

During the growth season, mortalities in the holding depots can be high due to problems caused by the physiological changes related to moulting. Normally, there are periods during crayfish season, when crayfish do not survive stressful conditions of holding due to proximity of moult, either premoult or post-moult [14]. These losses could cause problems in addition to holding conditions, i.e., lowered water quality in tanks or aggressive behaviour of other crayfish may elevate mortality of the most vulnerable crayfish. Crayfish in this physiological stage should not be taken into holding facilities.

A novel method for holding of crayfish for longer periods of time, called CrayShower, was first developed on Western Australia by Marron Farmer's Association during the 1990s. The method is based on combination of minimum water usage and cool temperature. Crayfish are held out water in misty air and cool temperature conditions in plastic containers for periods from days to several weeks. The method has been recently further developed in Finland and is currently used by some commercial crayfish trappers and traders (**Figure 6**) [33].

Figure 6. CrayShower with stacked plastic containers for holding of crayfish and a close-up of the water sprinkler system (small picture).

Crayfish can be held under these conditions for several months without significant losses in the number or quality of crayfish. Tasting tests indicated that crayfish held for 9 months in the CrayShower were of equal quality (test and texture of flesh) compared to those been held in tanks and given food for the same period. The survival among the CrayShower-held crayfish was significantly higher and stable throughout the holding period than tank-held crayfish. Thus, the CrayShower system, being slightly more expensive to build, would be more economical to utilise than the conventional holding in tanks. The benefits of the CrayShower would be less water utilised, higher survival of crayfish and decrease in workload.

The commercial value of crayfish benefits from purging [14], which normally means gut evacuation and maybe general cleaning of the carapace from loose debris and solid particles. During communal holding in tanks, with the catch fully submerged, gut evacuation happens normally within 24 hours, depending on the water temperature in the holding tanks [34]. This excrete also affects water quality, and holding tanks require constant water exchange. In CrayShower, the gut evacuation can take up to 2 weeks [35]. Gut evacuation improves crayfish survival during further transport to markets, and it makes crayfish more desirable gourmet food when processed and served, for obvious reasons.

Some crayfish wholesalers or processers demand clean crayfish, i.e., free of solid particle matter or discoloration, which might require cleaning of crayfish or sorting the catch by cleanliness. Normally, crayfish collect dirt on the carapace and become exceedingly dirty towards the end of the moult cycle especially late in the trapping season or when water temperatures remain low and moulting is infrequent. During warm summers or generally warm conditions with frequent moulting, on the other hand, crayfish are normally clean and, when cooked, display a nicely even red colour with no discoloration.

Crayfish should be gently packed in specific foam boxes for the transport to the markets [23]. There should not be more than two layers of crayfish with moisture and cooling provided inside the foam box. Quite often, shallow foam boxes with less than 15 cm in height, which are routinely used in fish transport, are convenient as they can be stacked easily or can be equipped with lids (**Figure 7**). No more than two layers of crayfish inside the box ensures that crayfish can be moving around, if needed, and the pressure from crayfish would not create circumstances which would suffocate those crayfish on the bottom. Cooling can be provided in the form of ice on the bottom of the foam boxes or cooling units [10], as long as it is ensured that crayfish would not be in direct contact with frozen solid items. It should be taken care of that the water level inside the foam box is low so that crayfish are not submerged and thus capable of gas exchange during transport. There is no need for ventilation holes in the foam box, as there is always enough oxygen in the air for crayfish and the content of metabolic gases created by gas exchange does not reach levels harmful for crayfish. The ventilation actually worsens the conditions during the transport, because it may cause elevation of the temperature due to loss of insulation and it may result in loss of moisture.

Figure 7. An example of the foam boxes suitable for the transport of crayfish on land. The corrugated bottom structure allows for maintaining moist conditions without crayfish being submerged.

6. Code of practice: a case study

We will present here a synopsis of the post-harvest handling code of practice for freshwater crayfish. This is based on this short overview with detailed background information given in the above chapters. These basic principles allow for maximum number of crayfish reaching markets. These principles can easily be applied to recreational trapping practices, too. We will not go into exhausting details here in order to allow for imaginative application of the recommendations. The following basic actions should ensure minimum stress for the catch and maximum benefit for crayfish trapper:

1. Sort crayfish on board according to the commercial requirements, and store them in cool environment, away from the elements, as soon as the traps have been pulled.

2. Aim for minimal handling of crayfish both initially and throughout the post-harvest processes.

3. Allow the crayfish transport conditions with a steady moist and cool environment and total isolation from the outside elements.

4. During the boat and road transport, ensure minimum changes in the ambient conditions and a low level of physical disturbance.

5. In the holding depots:

   a. Provide either cool water tanks with shade or specific holding conditions, such as CrayShower or similar. If stored for longer periods in tanks, provide hides and ensure sufficient water exchange.

   b. Sort crayfish according to the market requirements before putting them into the storage systems to avoid repeated handling.

6. Pack crayfish in shallow foam boxes for transport to markets, with cooling and moisture provided.

7. Throughout all stages of the post-harvest handling, remove weak, moribund and dead crayfish from the stock immediately.

Acknowledgements

Thanks to the crayfishermen, including Vesa Tiitinen, Esko Toikka and Timo Myllynen, trapping in Lake Saimaa (Finland), for showing their stamina against the elements of the Finnish summer and challenges of trapping crayfish. They inspired our development of post-harvest handling practices for Fennoscandian conditions.

Author details

Japo Jussila[1]*, Ravi Fotedar[2] and Lennart Edsman[3]

*Address all correspondence to: japo.jussila@uef.fi

1 Department of Environmental and Biological Sciences, University of Eastern Finland, Kuopio, Finland

2 Department of Environment and Agriculture, Curtin University, Bentley Campus, Western Australia, Australia

3 Department of Aquatic Resources, Swedish University of Agricultural Sciences, Sweden

References

[1] Lehtonen JUE. Kansanomainen ravustus ja rapujen hyväksikäyttö Suomessa. Helsinki: Oy Weilin+Göös Ab; 1975. p. 159. (in Finnish)

[2] Jussila J, Tiitinen V, Fotedar R, Kokko H. A simple and efficient cooling method for post-harvest transport of the commercial crayfish catch. Freshwater Crayfish. 2013;**19**(1):15-19. DOI: 10.5869/fc.2013.v19.015

[3] Jussila J, Mannonen A. Crayfisheries in Finland, a short overview. Bulletin Français de la Pêche et de la Pisciculture. 2004;**372-373**:263-273. DOI: 10.1051/kmae:2004001

[4] Jussila J, Mannonen A. Lisää hanaa. RapuSuihku -hanke, loppuraportti. Asikkala, Suomi: Raputietokeskus, Koulutuskeskus Salpaus; 2007. p. 8. (in Finnish)

[5] Swahn JÖ. The cultural history of crayfish. Bulletin Français de la Pêche et de la Pisciculture. 2004;**372-373**:243-251

[6] Taugbøl T. Exploitation is a prerequisite for conservation of *Astacus astacus*. Bulletin Français de la Pêche et de la Pisciculture. 2004;**372-373**:275-279. DOI: 10.1051/kmae:2004002

[7] Edsman L. Flodkräftan – värd att vårda. In: Blank S, Svensson M, editors. Artinriktad naturvård. Uppsala: Artdatabanken SLU; 2013. pp. 83-87. (in Swedish)

[8] Edsman L, Schröder S. Åtgärdsprogram för Flodkräfta 2008-2013 (*Astacus astacus*). Action plan for conservation of the noble crayfish. Fiskeriverket och Naturvårdsverket. Rapport 5955, Stockholm; 2009. p. 67. (in Swedish with English summary)

[9] Neill DM. Ensuring crustacean product quality in the post-harvest phase. Journal of Invertebrate Pathology. 2012;**110**:267-275. DOI: 10.1016/j.jip.2012.03.009

[10] Paterson DB, Spanoghe PT. Stress indicators in marine decapod crustaceans, with particular reference of western rock lobsters (*Panulirus cygnus*) during commercial handling. Marine and Freshwater Research. 1997;**48**:829-834

[11] Jussila J, Maguire I, Kokko H, Makkonen J. Chaos and adaptation in the pathogen-host relationship in relation to the conservation. The case of the crayfish plague and the noble crayfish. In: Kawai T, Faulkes Z, Scholtz G, editors. Freshwater Crayfish. A Global Overview. Warsaw: CRC Press; 2015. pp. 246-274. DOI: 10.1201/b18723-15

[12] Fotedar S, Evans LH. Health management during handling and live transport of crustaceans: A review. Journal of Invertebrate Pathology. 2011;**106**:143-152. DOI: 10.1016/j.jip.2010.09.011

[13] Paterson BD, Spanoghe PT, Davidson GW, Hosking W, Nottingham S, Jussila J, et al. Predicting survival of western rock lobsters *Panulirus cygnus* using discriminant analysis of haemolymph parameters taken immediately following simulated handling treatments. New Zealand Journal of Marine and Freshwater Research. 2005;**39**:1129-1143. DOI: 10.1080/00288330.2005.9517380

[14] McClain WR. Assessment of depuration system and duration on gut evacuation rate and mortality of red swamp crawfish. Aquaculture. 2000;**186**(3-4):267-278. DOI: 10.1016/ S0044-8486(99)00377-4

[15] Jussila J, Paganin M, Mansfield S, Evans LH. On physiological responses, plasma glucose, total hemocyte counts and dehydration, of marron *Cherax tenuimanus* (Smith) to handling and transportation under simulated conditions. Freshwater Crayfish. 1999;**12**:154-167

[16] Jussila J. On the economics of crayfish trapping in Central Finland in 1989-90. Freshwater Crayfish. 1995;**8**:215-227

[17] Ackefors H. The positive effects of established crayfish introductions in Europe. In: Gherardi F, Holdich DM, editors. Crayfish in Europe as Alien Species. How to Make the Best of a Bad Situation. Rotterdam, Netherlands: A.A. Balkema; 1999. pp. 49-62

[18] Patullo BW, Baird HP, Macmillan DL. Altered aggression in different sized groups of crayfish supports a dynamic social behaviour model. Applied Animal Behaviour Science. 2009;**120**(3-4):231-237. DOI: 10.1016/j.applanim.2009.07.007

[19] Steele C, Skinner C, Alberstadt P, Antonelli J. SHORT COMMUNICATION: Importance of adequate shelters for crayfishes maintained in aquaria. Aquarium Science and Conservation. 1997;**1**(3):189-192. DOI: 10.1023/A:1018304205540

[20] Morrissy NM, Caputi N. Use of catchability equations for population estimation of marron, *Cherax tenuimanus* (Smith) (Decapoda: Parastacidae). Australian Journal of Marine and Freshwater Research. 1981;**32**:213-225. DOI: 10.1071/MF9810213

[21] Huner JV, editor. Freshwater Crayfish Aquaculture in North America, Europe, and Australia: Families Astacidae, Cambaridae, and Parastacidae. London, UK: CRC Press; 1994. p. 336

[22] Morrissy NM, Walker P, Fellows C, Moore W. An investigation of weight loss of marron (*Cherax tenuimanus*) in air during live transportation to market. Bernard Bowen Fisheries Institute, Western Australian Marine Research Laboratories. Fisheries Department of Western Australia, Fisheries Report. 1997;**99**:1-21

[23] Lorenzon S, Giulianini PG, Libralato S, Martinis M, Ferrero EA. Stress effect of two different transport systems on the physiological profiles of the crab *Cancer pagurus*. Aquaculture. 2008;**278**:156-163. DOI: 10.1016/j.aquaculture.2008.03.011

[24] Aydin H, Jussila J, Kokko H, Tiitinen V. Rapulaatikko - rei'illä vai ilman?. Suomen kalastuslehti. 2013;**120**(6):28-30. (in Finnish)

[25] Terchunian AV, Kunz NA, O'Dierno LJ, editors. Air Shipment of Live and Fresh Fish & Seafood Guidelines. A Manual on Preparing, Packing and Packaging Live and Fresh Fish & Seafood Air Shipments along with Customs and Inspection Guidelines for Six APEC Member Economies. 1st ed. Singapore: The APEC Secretariat; 1999. p. 190

[26] Jussila J, Jago J, Tsvetnenko E, Evans LH. Effects of handling or injury disturbance on total hemocyte counts in western rock lobster (*Panulirus cygnus* George). In: Evans LH, Jones

B, editors. Proceedings of the International Symposium on Lobster Health Management, Adelaide. 1st ed. Perth, Western Australia: Curtin University; 2001. pp. 52-62

[27] Jussila J, McBride S, Jago J, Evans LH. Hemolymph clotting time as an indicator of stress in western rock lobster (*Panulirus cygnus* George). Aquaculture. 2001;**199**:185-193. DOI: 10.1016/S0044-8486(00)00599-8

[28] Jussila J, Jago J, Tsvetnenko E, Dunstan B, Evans LH. Total and differential haemocyte counts in western rock lobster (*Panulirus cygnus* George) under post-harvest handling stress. Marine and Freshwater Research. 1997;**48**:863-867. DOI: 10.1071/MF97216

[29] Spanoghe PT. An investigation of the physiological and biochemical responses elicited by *Panulirus cygnus* to harvesting, holding and live transport [thesis]. Perth, Western Australia: Curtin University of Technology; 1997. p. 378

[30] Fotedar S, Evans LH, Jones B. Effect of holding duration on the immune system of western rock lobster, *Panulirus cygnus*. Comparative Biochemistry and Physiology A. 2006;**52**:1351-1355. DOI: 10.1016/j.cbpa.2006.01.010

[31] Jussila J, Mannonen A, Kilpinen K. Ravun laatu ja uudet tuotteet. Suomen kalastuslehti. 2009;**116**(7):22-24

[32] Jussila J. Notes on marron response to high temperature stress. ACWA News. 1995;**9**:27-29

[33] Jussila J, Tiitinen V. Suolen tyhjeneminen RapuSuihkussa. Loppuraportti. Kuopio: Saimaan rapu -hanke; 2011. p. 5 (in Finnish)

[34] Rosén N. Svenskt fiskelexikon. Stockholm, Sweden: Nordiska Uppslagsböcker; 1956. p. 704

[35] Anon. Hans Kungliga Majestäts befallningshavandes femårsberättelser; jämte Sammandrag för åren 1896-1900, Stockholms län, på Nådigaste befallning utarbetat och utgivet af Statistiska centralbyrån. Stockholm: SCB; 1901. p. 44 (in Swedish)

Permissions

All chapters in this book were first published in PH, by InTech Open; hereby published with permission under the Creative Commons Attribution License or equivalent. Every chapter published in this book has been scrutinized by our experts. Their significance has been extensively debated. The topics covered herein carry significant findings which will fuel the growth of the discipline. They may even be implemented as practical applications or may be referred to as a beginning point for another development.

The contributors of this book come from diverse backgrounds, making this book a truly international effort. This book will bring forth new frontiers with its revolutionizing research information and detailed analysis of the nascent developments around the world.

We would like to thank all the contributing authors for lending their expertise to make the book truly unique. They have played a crucial role in the development of this book. Without their invaluable contributions this book wouldn't have been possible. They have made vital efforts to compile up to date information on the varied aspects of this subject to make this book a valuable addition to the collection of many professionals and students.

This book was conceptualized with the vision of imparting up-to-date information and advanced data in this field. To ensure the same, a matchless editorial board was set up. Every individual on the board went through rigorous rounds of assessment to prove their worth. After which they invested a large part of their time researching and compiling the most relevant data for our readers.

The editorial board has been involved in producing this book since its inception. They have spent rigorous hours researching and exploring the diverse topics which have resulted in the successful publishing of this book. They have passed on their knowledge of decades through this book. To expedite this challenging task, the publisher supported the team at every step. A small team of assistant editors was also appointed to further simplify the editing procedure and attain best results for the readers.

Apart from the editorial board, the designing team has also invested a significant amount of their time in understanding the subject and creating the most relevant covers. They scrutinized every image to scout for the most suitable representation of the subject and create an appropriate cover for the book.

The publishing team has been an ardent support to the editorial, designing and production team. Their endless efforts to recruit the best for this project, has resulted in the accomplishment of this book. They are a veteran in the field of academics and their pool of knowledge is as vast as their experience in printing. Their expertise and guidance has proved useful at every step. Their uncompromising quality standards have made this book an exceptional effort. Their encouragement from time to time has been an inspiration for everyone.

The publisher and the editorial board hope that this book will prove to be a valuable piece of knowledge for researchers, students, practitioners and scholars across the globe.

List of Contributors

Tebien Federico Hahn
Chapingo Autonomous University, Texcoco, Mexico

Jason Sun and Rainer Künnemeyer
School of Engineering, University of Waikato, Hamilton, New Zealand
Dodd Walls Centre for Photonic and Quantum Technologies, New Zealand

Andrew McGlone
The New Zealand Institute for Plant and Food Research, Hamilton, New Zealand

Sandra Horvitz
Food Science and Engineering Faculty, Technical University of Ambato, Ambato, Ecuador

Diego A. Castellanos
Food Engineering Program, Agricultural University Foundation of Colombia, Bogotá, Colombia
Postharvest Laboratory, National University of Colombia, Bogotá, Colombia

Aníbal O. Herrera
Postharvest Laboratory, National University of Colombia, Bogotá, Colombia

Hatice Serdar and Serhat Usanmaz
European University of Lefke, Lefke, Cyprus

Afam I.O. Jideani, Tonna A. Anyasi, Elohor O. Udoro and Oluwatoyin O. Onipe
Department of Food Science and Technology, School of Agriculture, University of Venda, Thohoyandou, Limpopo Province, South Africa

Godwin R.A. Mchau
Department of Horticultural Sciences, School of Agriculture, University of Venda, Thohoyandou, Limpopo Province, South Africa

Francisco Artés-Hernández, Ginés Benito Martínez-Hernández, Encarna Aguayo and Francisco Artés
Postharvest and Refrigeration Group, Department of Food Engineering, Universidad Politécnica de Cartagena, Cartagena, Spain

Perla A. Gómez
Institute of Plant Biotechnology, Universidad Politécnica de Cartagena, Edificio I+D+i, Cartagena, Spain

Puran Bridgemohan
Faculty of Biosciences, Agriculture and Food Technologies, The University of Trinidad and Tobago, Trinidad and Tobago

Wendy-Ann P. Isaac
Department of Food Production, Faculty of Food and Agriculture, The University of the West Indies, St. Augustine, Trinidad and Tobago

Japo Jussila
Department of Environmental and Biological Sciences, University of Eastern Finland, Kuopio, Finland

Ravi Fotedar
Department of Environment and Agriculture, Curtin University, Bentley Campus, Western Australia, Australia

Lennart Edsman
Department of Aquatic Resources, Swedish University of Agricultural Sciences, Sweden

Index

www.ingramcontent.com/pod-product-compliance
Lightning Source LLC
Chambersburg PA
CBHW062005190326
41458CB00009B/2974